The inferred positions of global paleoshorelines through the 240 million years of the Mesozoic and Cenozoic are presented within this atlas. Thirty-one maps, generally corresponding to stratigraphic stages, provide a snapshot of the continents and their shorelines at approximately 8 million year intervals. The maps provide a representation of the gross changes in the distribution of land and sea throughout the Mesozoic and Cenozoic plotted on Mollweide projections of paleocontinental reconstruction. They do not distinguish between well and poorly defined shorelines, but the information sources are set out in a bibliography numbering more than 2000 primary paleographic references. Work up to 1980 was used in the initial compilation, and the final maps use additional unreferenced and unpublished data up to about 1987.

This is a unique global compilation that presents the first attempt at delineating global shorelines at stage level.

Atlas of Mesozoic and Cenozoic Coastlines

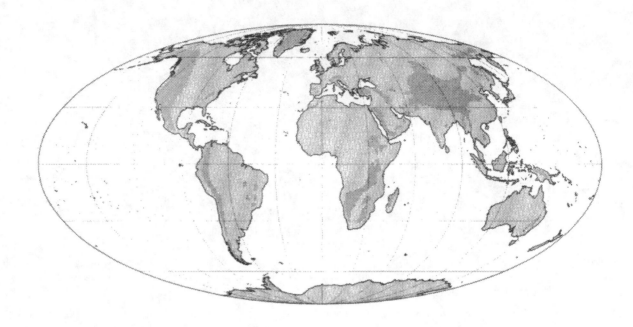

Present-day coastline map.

ATLAS OF MESOZOIC AND CENOZOIC COASTLINES

ALAN G. SMITH

Department of Earth Sciences, Cambridge

DAVID G. SMITH

Basin Research Group, Petroconsultants (UK) Ltd

and

BRIAN M. FUNNELL

School of Environmental Sciences, University of East Anglia

PUBLISHED BY THE PRESS SYNDICATE OF THE UNIVERSITY OF CAMBRIDGE
The Pitt Building, Trumpington Street, Cambridge, United Kingdom

CAMBRIDGE UNIVERSITY PRESS
The Edinburgh Building, Cambridge CB2 2RU, UK
40 West 20th Street, New York NY 10011–4211, USA
477 Williamstown Road, Port Melbourne, VIC 3207, Australia
Ruiz de Alarcón 13, 28014 Madrid, Spain
Dock House, The Waterfront, Cape Town 8001, South Africa

http://www.cambridge.org

First published 1994
First paperback edition 2004

A catalogue record for this book is available from the British Library

Library of Congress cataloguing in publication data

Smith, A. Gilbert (Alan Gilbert), 1937–
Atlas of Mesozoic and Cenozoic coastlines / Alan G. Smith, David
G. Smith, and Brian M. Funnell
p.   cm.
Includes bibliographical references.
ISBN 0 521 45155 8 hardback
1. Geology, Stratigraphic – Mesozoic – Maps.   2. Geology,
Stratigraphic – Cenozoic – Maps.   3. Coasts – Maps. 4. Paleogeography –
– Maps.   I. Smith, David G. (David Graham)   II. Funnell, B. M.
III. Title.
G1046.C5S6   1994 <G&M>
551.7´6´0223 – dc20   93-28047 CIP MAP

ISBN 0 521 45155 8 hardback
ISBN 0 521 60287 4 paperback

Contents

Contents

Preface

The maps in this atlas originated in concurrent Extramural Research Awards by British Petroleum from 1981 to 1985 to the Department of Earth Sciences at the University of Cambridge and the School of Environmental Sciences of the University of East Anglia. The group at the University of East Anglia (B. M. Funnell, R. V. Tyson and T. Austin) compiled paleocoastlines and paleofacies on present-day base maps of the individual continents, from published paleogeographic maps and other data. The Cambridge group (A. G. Smith, R. A. Livermore, L. A. Rush, J. L. Tarry) prepared the present-day coastlines, bathymetry, geographic grids and sutures, assessed the data for making global reconstructions and developed the ATLAS map-plotting and paleocontinental reconstruction software and associated databases. The Global Paleoreconstruction Group at BP Research (initially under I. R. Hoskin, then C. P. Summerhayes, then D. G. Smith), drawing on the work of both university groups, developed their own GLOBE computer reconstruction software with which regional plate tectonic studies were conducted and global reconstruction atlases were developed for BP's internal use. The paleocoastlines in the present atlas were initially compiled as an internal report. The BP group also supported the Paleogeographic Atlas Project at the University of Chicago, under Alfred Ziegler, and their global data compilations were also drawn upon at various stages of this work.

The maps presented here represent very many man-years of work, and we gratefully acknowledge the following for their input at various stages: Peter Llewellyn, Richard Field, Ian Hoskin and Andrew Fleet, who were responsible for setting the project up in the former Stratigraphy Branch of BP Research's Exploration and Production Division; David Burwell, Raymond Jackson and Brian Biart, who wrote and maintained the GLOBE continental reconstruction and database software; Neil Parkinson, Helen Turnell, Michael Daly and Martin Heffernan, who were all closely involved with the regional and global studies with BP's Global Paleoreconstruction research group that contributed to the reconstructions herein. Several generations of geology students were employed as temporary technical assistants at BP Research and made valuable contributions. BP Exploration, through Michael Daly, provided finance for the editing in Cambridge of the bibliography and maps by Clare Glover and Graham Cocksworth. Throughout the project there was considerable help from secretarial staff in all three organizations. Finally, we are grateful to BP International plc for permission to publish these maps.

_ l _
Introduction

This atlas is one outcome of a substantial research programme that was conducted under the sponsorship of British Petroleum International. BP established concurrent research projects over a four-year period, 1981–1985, at the Universities of Cambridge (Department of Earth Sciences) and East Anglia (School of Environmental Sciences). The aim was to produce up-to-date computer-drawn maps on which were plotted global paleogeographic data at the stratigraphic resolution of one map for each Stage for most of Mesozoic and Cenozoic time.

The Cambridge project, under the direction of Alan Smith, drew on that department's expertise, dating back to the work of Bullard, Everett and Smith (1965) and including earlier collections of maps by Smith, Briden and Drewry (1973), and Smith, Hurley and Briden (1981). They developed the ATLAS reconstruction software, initially on a mainframe computer but more recently on a PC, and prepared the large amounts of digitized data that make up the continental fragments, including present-day coastlines, bathymetric contours, fragment boundaries, and intra-continental sutures.

The University of East Anglia project was responsible for the paleogeographic and paleoenvironmental data to be used on the resulting global reconstructions. R. V. Tyson conducted a systematic search of the recent geological literature, back to a cut-off date of about 1960. Rather than attempting to collect primary data, the search was for data already compiled into paleogeographic maps, whether at local, regional or global scale. As far as possible, the maps were sought relating to a stratigraphic interval no longer than a Stage. The search resulted in maps from about 2000 sources, which are listed in the Bibliography. The Stage was chosen as the appropriate stratigraphic unit because it is the smallest chronostratigraphic division that can usually be recognized globally. The information from these maps was compiled, a Stage at a time, on to regional maps, generally covering a continent each, and using a present-day geographic base. The 31 resulting maps are the first known attempt to draw a continuous series of global maps with such a fine resolution. In some cases, especially for the Triassic, adjacent Stages have had to be combined because of lack of data, but the average duration of the stratigraphic units for the whole series is about 8 million years (Ma) (31 maps covering 245 Ma).

Meanwhile, the research group at BP developed their own paleoreconstruction software (GLOBE), with an associated database from which lithofacies data could be plotted directly on to paleocontinental reconstructions. This system was used to conduct a series of regional plate tectonic studies for BP Exploration, on regions that included the Arctic (Smith 1987), northern Australia/Papua New Guinea, the Caribbean, Southeast Asia (Daly *et al.* 1991), northwest Europe, East Africa/Arabia, the South Atlantic and New Zealand. Originally for internal company use, the results of all of those studies contributed to the development of BP's in-house global reconstructions and thus to the maps in this Atlas. Although initially based on the Cambridge reconstructions, BP's global reconstructions inevitably came to diverge in some respects from those of ATLAS, and the reconciliation of the two is described below.

The final stage in BP's involvement with this project, postdating the conclusion of the sponsored research in both universities, was the merging of the topographic data on East Anglia's continent-by-continent compilations, first on to present-day global base-maps and then on to reconstructions appropriate to the age of the paleogeographic information. The resulting map series was transferred to Cambridge for reconciliation with the ATLAS reconstructions, and computer-plotting of the maps as they appear in this Atlas.

Global shoreline maps for the Cretaceous, based on the original 1986 versions, were published in 1990 (Funnell 1990). More detailed Cretaceous shoreline maps for Europe only, based on the 1986 continent-by-continent facies and shoreline maps, were first published in 1987 and republished in 1990 (Tyson and Funnell 1987; Funnell 1990).

— 2 —
Global reconstructions

2.1 Making a continental reassembly

The method of making global continental reconstructions is now routine. It has been fully described elsewhere (e.g. Smith, Briden and Drewry, 1973; Smith, Hurley and Briden 1981) and only an outline is given here.

The first step is to subdivide the present-day world into the continental fragments that existed at the time concerned. We use the word 'fragment' because the various pieces to be reassembled do not all coincide neatly with individual continents. Each continental fragment consists essentially of a digitized outline, each point on which must be moved to a different position for each new reconstruction. As well as the outline, other digitized data may also be associated with the fragment, and moved with the outline. Thus, we normally include the present-day coastline, not for any paleogeographic significance, but simply in order that the fragment may be readily identified. Continental fragment boundaries rarely coincide with present-day coastlines, but are more usually located at continent–ocean boundaries or at intra-continental sutures.

The primary source of information for making Mesozoic and Cenozoic reconstructions is the ocean-floor magnetic anomaly record. All magnetic chrons recorded in the magnetic anomalies of the Atlantic and Indian Oceans are fitted together as well as possible. The information required for the computer to transform each pair of present-day coordinates into the corresponding pair of paleo latitude–longitude coordinates is expressed as an angle of rotation and the coordinates of the pole of rotation, using Euler's principle (see Cox and Hart 1986, for a comprehensive account of the method). We have used only published finite rotations in this work (Table 1), because the effort of reexamining the original ocean-floor anomaly data is a major research project in itself. The precision with which anomaly matches can be made has been greatly increased by using SEASAT data (e.g. Cande, LaBrecque and Haxby 1988). Modifications to the geologic time-scale have also increased the precision of the reconstructions (Harland *et al.* 1990). The increase in precision that could be obtained by using the most recent revision of the Late Cretaceous to Recent magnetic polarity time-scale (Cande and Kent 1992) is not significant on the scale of the maps.

Table 1. *Sources of rotation data*

Region	Reference
Atlantic Ocean data	
S. America to Africa	
C4N to C34N	Cande *et al.* 1988
'M0' (Late Aptian) to closure	Klitgord and Schouten 1990
N. America to Greenland	
C13N	Srivastava and Tapscott 1986
C21N-'M0'	Roest and Srivastava 1989
Pre-closure	Smith and Livermore 1991
N. America to Europe	
C5 to CRT	Srivastava and Tapscott 1986
Closure	Bullard *et al.* 1965
Indian Ocean data	
Africa to Somalia	
C5	McKenzie, Davies and Molnar 1970; Le Pichon and Francheteau 1978
Closure	Bullard *et al.* 1965
Madagascar to Somalia	Segoufin and Patriat 1981
Antarctica to Madagascar	
C3N–C5N,C7N–C8N,C21N	Cande and Leslie 1986
C6N	Sclater *et al.* 1981
C13, C22N, C34N	Fisher and Sclater 1983
C28N	Lawver, Sclater and Meinke 1985
Australia to Antarctica	
C5N–C13N, C22N, C31N	Stock and Molnar 1982
C18N, C24N, C34N	Mutter *et al.* 1985
Closure	Weissel *et al.* 1977
India to Australia	
C22N–C28N	Norton and Sclater 1979
C34	Fisher and Sclater 1983
M1R and closure	Norton and Sclater 1979

Where the anomaly data do not coincide in time with the age required for a particular reconstruction, the fit is interpolated between the finite rotations of the times closest to it. This procedure gives continental reassemblies for all the major continents back to the age of the oldest ocean-floor (mid-Jurassic). The continental reassembly of earliest Jurassic time is assumed to have been the classic Pangea obtained by fitting the continental edges together, but the Triassic maps have been modified to take into account Triassic extension and motion along a postulated transform zone (Smith and Livermore, 1991) that makes the Early Triassic reconstruction very similar to that of Van der Voo and French (1974) as modified by Walper (1980).

Where triple junctions have existed between three plates, such as during part of the opening of the North Atlantic Ocean between North America, Greenland and Europe, the reassembly can be made in two ways. For example, Greenland can be reassembled relative to Europe, and the Europe–Greenland cluster can be moved to North America. Alternatively, Greenland and Europe can be independently brought to North America. The routes chosen were considered at the time to have the best-known rotations but in the future new data may suggest a different route.

2.2 Sources of finite rotation data

Table 1 summarizes the sources of the data used for reconstructing pairs of continents.

2.3 Areas affected by orogenesis

The major uncertainties in the maps are the positions of those continental fragments that have undergone compressional deformation at a later time than that for which the map is required. For example, all continental fragments in the Alpine–Himalayan chain have been affected by Cenozoic deformation. Their positions in pre-Cenozoic reassemblies are the least well known of all the continental fragments.

There have been no major new insights into how such fragments can be repositioned in a reassembly, but paleomagnetic data from these and so-called exotic or suspect terranes will require significant shifts in the positions of some of these fragments. But because the areas affected by Mesozoic and Cenozoic deformation are relatively small, errors in the reconstruction of these fragments do not significantly affect the global picture. Despite an increase in paleomagnetic and other relevant data from Southeast Asia it has been arbitrarily kept attached to Asia in its present-day position and configuration. Paleomagnetic data that became available after the compilation of the maps suggest significant modifications to the positions of China and neighbouring regions to those shown on the Jurassic and Triassic maps (Enkin *et al.* 1992).

The maps show some continental fragments such as northern and southern Tibet sweeping across the Tethys. This representation is largely schematic; these fragments probably did traverse the Tethys, but their shapes and the timing of their migration are illustrative rather than definitive. No attempt has been made to document the data on which the positions of these fragments and other fragments caught up in the Cenozoic–Mesozoic orogenic belts are based. Differences in the positions of these fragments will not alter the global area covered by sea but only the distribution of that area.

2.4 Orienting a reassembly

The reassembled continents must then be oriented so that the global mean paleomagnetic pole of the time concerned coincides with the map projection pole. The availability of paleomagnetic catalogues (Piper 1988) and machine-readable paleomagnetic databases (Lock and McElhinny 1991) has led to an increase in the number of poles that may used for this purpose.

Reliable paleomagnetic pole positions for the tectonically stable areas are rotated back to a reference frame – in this case the reference fragment used is Africa. For Mesozoic and Cenozoic time there is no ambiguity about whether the pole is the North or South pole. A window of 30 Ma has been applied to the paleomagnetic data and the global mean pole on Africa has been determined at 20 Ma intervals. Thus the mean paleomagnetic pole for a map, say, for 100 Ma has been found by averaging all poles lying in the 85–115 Ma age range. The mean age of the data will in general differ from the mean age of the window and the actual pole position at 100 Ma is interpolated between two bracketing ages, e.g. between 97 Ma and 118 Ma. The global mean pole for each stage is the interpolated position of the pole whose age is the mean age of the stratigraphic stage concerned. The estimate of the Earth's spin axis on the reassembly, i.e. its polar axis, is taken as the global mean pole. A map is then made by projecting the reassembly as a global map using the global mean paleomagnetic pole as the geographic pole for the time concerned. The new databases have decreased the uncertainty of the global paleomagnetic poles for Mesozoic and Cenozoic reassemblies but have not greatly changed the positions of the mean poles. Alpha-95 values – estimates of the uncertainty in pole position and therefore of paleolatitude – rarely exceed 5° and are generally much less.

Despite the availability of all these new data, comparisons of the present maps with previously published maps (e.g. Smith *et al.* 1981) show that the new maps are refinements of previous reconstructions rather than representing radically new continental distributions. It is clear that any new data are most unlikely to alter the positions of any of the major continents in any significant way, particularly on the scale adopted.

2.5 Choice of map projection

Projecting a sphere on to a plane always poses a problem. Since the intention is to show a global view then Mercator's projection cannot be used. Originally, it was intended to use two projections, one highlighting the equatorial regions and another for the two polar regions. However, this would have tripled the number of maps. The atlas uses Mollweide's projection, an equal-area projection commonly used to show the whole globe. Because the maps are computer generated, it is in practice a trivial matter to generate maps on a variety of other projections. All maps depict the present-day coastline as well as the paleocoastline. The present-day coastline is shown so that the continental fragments can be readily recognized. The geographic grid that has been superposed on the maps is centred on the mean paleomagnetic poles of the time concerned. Latitude and longitude lines are spaced at 30° intervals.

_ 3 _
Biostratigraphic time and magnetic polarity scales

3.1 General statement

The biostratigraphic subdivisions and numerical ages in Ma assigned to them are those of Harland et al. (1990). Compared with other time-scales, this scale gives one of the more extreme (older) values for the Jurassic/Cretaceous boundary (at 146 Ma). Modifications to parts of the Cenozoic scale have been made (Berggren *et al.*, 1991) and others suggested to the Mesozoic scale (Gradstein *et al.*, 1993). The magnetic polarity scale and the ages assigned to polarity changes are also taken from Harland *et al.* (1990).

The Cenozoic magnetic polarity scale is also under revision (Cande and Kent 1992). The revisions are of two types: those that change the relative spacing of the polarity changes and those that modify their numerical ages. Changes in relative spacing, if confirmed, will affect all subsequent magnetic polarity scales; modifications to the numerical ages are likely to be matters of discussion and significant refinement for some time to come.

3.2 Stratigraphic subdivisions used

Thirty one maps have been drawn for Mesozoic and Cenozoic time. Their stratigraphic age ranges, alphabetic abbreviations and numerical ages (Table 2) are taken from Harland *et al.* (1990). Where two or more Stages are covered by a single map, e.g. the Messinian–Tortonian, making up the Late Miocene, the younger Stage is listed first. For the older Jurassic and part of the Triassic, the numerical ages have been rounded to the nearest 5 Ma. The Bathonian and Aalenian maps have been excluded for lack of data. However, the Callovian and Bajocian maps, which bracket the Bathonian, are similar to each other as are the Bajocian and Toarcian maps, which bracket the Aalenian. The unpublished University of East Anglia Bathonian and Aalenian shoreline maps show them to be intermediate between the bracketing Stages.

Table 2. *Stratigraphic and numerical time-scale*

Stratigraphic age	Abbreviation	Numerical age (Ma)
Tertiary		
Neogene		
Pliocene	Pli	5
Late Miocene (Messinian–Tortonian)	Mio3	10
Middle Miocene (Serravallian–Langhian)	Mio2	12
Early Miocene (Burdigalian–Aquitanian)	Mio1	20
Paleogene		
Oligocene	Oli	30
Late Eocene (Priabonian)	Eoc3	37
Middle Eocene (Bartonian–Lutetian)	Eoc2	45
Early Eocene (Ypresian)	Eoc1	53
Paleocene (Thanetian–Danian)	Pal	60
Cretaceous		
Late Cretaceous		
Maastrichtian	Maa	70
Campanian	Cmp	80
Santonian	San	85
Coniacian	Con	88
Turonian	Tur	90
Cenomanian	Cen	95
Early Cretaceous		
Albian	Alb	105
Aptian	Apt	115
Barremian–Hauterivian	Brm–Hau	125
Valanginian–Berriasian	Ber–Vlg	138
Jurassic		
Late Jurassic (= Malm)		
Tithonian	Tth	148
Kimmeridgian	Kim	153
Oxfordian	Oxf	155
Middle Jurassic (= Dogger)		
Callovian	Clv	160
Bathonian	Bth	Map not made
Bajocian	Baj	170
Aalenian	Aal	Map not made
Early Jurassic (= Lias)		
Toarcian	Toa	180
Pliensbachian	Plb	190
Sinemurian	Sin	200
Hettangian	Het	205
Triassic		
Late Triassic (Rhaetian–Norian–Carnian)	Tr3	220
Middle Triassic (Ladinian–Anisian)	Tr2	238
Early Triassic (= Scythian)	Tr1	245

Abbreviations from Harland *et al.* (1990). Stratigraphic ranges from Harland *et al.* (1990) but ages do not necessarily correspond to the mid-points of each range.

_ 4 _
Plotting paleogeographic data

4.1 Initial map preparation

The paleogeographic data were compiled initially on base maps on a scale of about 1:10 m. At such a scale it was possible to distinguish a number of marine and non-marine facies. However, information at this level of detail was routinely available (in the public domain) only for western Europe and North America. These unpublished detailed facies maps are the basis for the paleocoastline maps in this Atlas.

The total Bibliography of sources for the information summarized in the shoreline maps given on pp. 55–99 provides an essential pathway into more detailed local or regional information. It is important to note, however, that this information covers only the period up to 1985. More recent literature is not included or referred to.

In addition to the literature compiled by the University of East Anglia, BP had access to the global reconstructions of the Paleogeographic Atlas Project of the Department of Geophysics, University of Chicago (A. M. Ziegler, personal communication), a version of which has been published (Ziegler, Scotese and Barrett 1982). This series comprised some 16 reconstructions, each representing a single Stage, for which data had been compiled from the literature at a density of up to one datapoint for each one degree latitude/longitude square.

Areas of higher relief for the time concerned are shown schematically on each map. These are generalized from BP's database, which is a modified version of the data collected by F. Ziegler's group at the University of Chicago.

The global shoreline maps that form the basis of this atlas were compiled by staff and vacation students over a two-year period at the BP Research Centre. The compilation was initially on a present-day geographic base, with a map for each Stage from the base of the Jurassic to the Pliocene. The University of East Anglia compilation did not include the Triassic, and maps for each of the Early, Mid and Late Triassic were added to the series at this stage.

4.2 Transfer to global reconstructions

The shorelines were then transferred to global continental reconstructions for each Stage, using reconstructions generated by BP's GLOBE system. Because of the termination of the Global Paleoreconstruction project in BP, it became necessary to transfer the paleocoastlines to ATLAS for final plotting for this publication. However, by this time the detailed rotations that specify the paleocontinental positions were significantly different between ATLAS and GLOBE. Also, because the reconstructed positions of the shorelines were digitized, but their present-day positions were not, it was necessary to go through a rather cumbersome procedure at this point. The shorelines had to be divided into segments corresponding to the continental fragments used in GLOBE, then back-rotated using GLOBE's rotation data to recover the original position of the shorelines on present-day maps. They were then rotated again using the ATLAS rotations, and tidied up before generating the plots presented in this volume. Difficulties were experienced with back-rotating the data for only two maps – the Valanginian and Berriasian map, and the Barremian and Hauterivian map. The shorelines on both these maps have been redigitized from their present-day positions rather than back-rotating the old reconstructed data.

4.3 Estimation of paleocoastline positions

The shoreline is the paleogeographic element that is perhaps most commonly singled out on maps of this kind. The scientific justification for selecting this particular feature is not obvious, because the coastline is a notably ephemeral feature – it may migrate backwards and forwards across large areas of a continent with considerable rapidity. However, it provides geologists with a very sensitive indicator of changes in base level, particularly in areas of low gradient where a small relative rise or fall in sea level will cause a correspondingly large movement of the coastline. The level of interest in the degree of control exercised by changing sea-levels on patterns of sedimentation and stratigraphic packaging is currently high, and maps of the present time are a potentially important contribution to this discussion (see below).

Ideally, placing a paleocoastline on a map is no more than drawing the boundary between the extent of marine sedimentation on one side of the line and continental on the other. Only in the most intensively researched parts of the globe, principally in western Europe and North America, is adequate information available. At its best this may involve well-dated sequences of marginal marine (paralic) facies where the saltwater/freshwater or marine/terrestrial interfaces can be defined in actual sedimentary deposits. More commonly only terrestrial-freshwater or marine-saltwater deposits are present, and the shoreline is defined from only one side. Particularly for periods of major marine transgression the marginal marine deposits may be subsequently eroded away, and their preserved extent is then only an indication of the minimum extent of the transgression. In these circumstances variations in the apparent depth of deposition of the marine deposits have been interpreted as indicators of the relative extent of transgressions and commensurate shoreline displacement. Further consideration of the

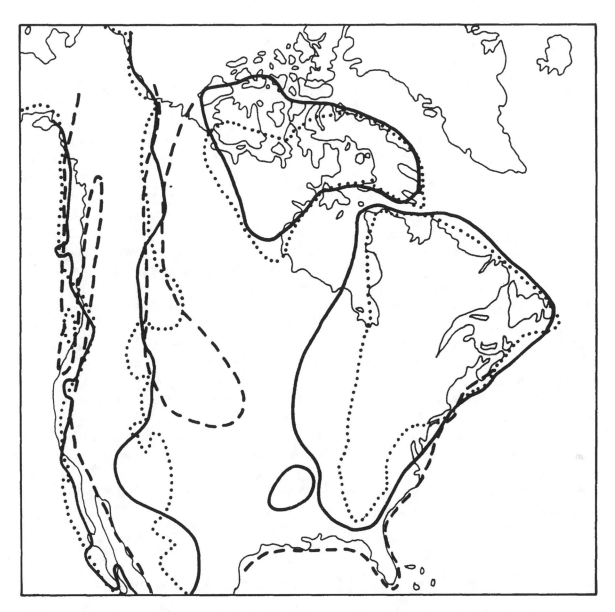

Figure 1. Three Maastrichtian coastlines for North America. The full line is from Map 10, this atlas; the dotted line is from E. C. Kauffman and D. A. Beeson (unpublished results) and the dashed line from Funnell (1990). Lambert equal area map with a central meridian at 90° W. See text for discussion.

limitations of global and regional maps in respect of paleocoastline and facies distributions have been given by Tyson and Funnell (1987) and Funnell (1990).

In practice the low density of the available data dictates that it is very rarely so simple, and paleocoastlines at the global scale must be regarded as very speculative. For many geological Stages, there are simply no data at all from many parts of the globe. Under these circumstances, it may be possible to make reasonable estimates of the likely shoreline position from its position in preceding or succeeding Stages. In the less explored or researched parts of the globe such interpolation may extend over many Stages and real control over the shoreline position is very tenuous. An example is Antarctica, for which most of the maps simply assume that the continent was elevated for most of Mesozoic and Cenozoic time (except where marine deposits are known to be present), and we therefore show a paleocoastline that approximates to that of the present day.

As examples of the variation in interpretation that is possible between different research groups, the Cretaceous shoreline maps of Tyson and Funnell (1987) and Funnell (1990) can be compared directly with the maps in this atlas which were prepared by different researchers but from very much the same data (Figure 1). They may also be compared with a third set of maps for North America (E. C. Kauffman and D. A. Beeson, unpublished data). The philosophy adopted in compiling our maps was to illustrate, in cases of diverging interpretations for the same Stage, the maximum amount of marine transgression. The Maastrichtian maps are of particular interest because they show the greatest divergence. Funnell's (1990) map shows no through-going north-south seaway east of the Rocky Mountains. By contrast, the map in this atlas and that of Kauffman and Beeson show a wide seaway forking into two, with one branch extending to Hudson's Bay and beyond. Though broadly similar, the paleocoastlines of these two maps differ in detail. By the Paleocene, the western seaway had disappeared. The differences in all three maps reflect mostly the duration and age of the time-slice used. Kauffman and Beeson's map is for a very short time interval – the *Baculites clinolobatus* chron in division R9 of the Bearpaw cycle – whereas the other two are representative of different parts of the Maastrichtian. Thus Funnell's (1990) map is probably representative of a later part of the Maastrichtian than the map in this Atlas. The latter probably represents the average Maastrichtian shoreline during the earlier Maastrichtian. All three maps are therefore acceptable on their own terms, but together they illustrate the difficulties of drawing and evaluating individual paleocoastline maps, particularly when large variations take place within a Stage.

4.4 An illustration of the use of the Bibliography

To give some idea of the number and distribution of the references used to compile a single global map, we list here the references used for the Albian Stage map.

In preparing the original shoreline and facies maps, data were generally consolidated on to regional base maps. These regions comprised: Arctic (including the USSR), North America (including the Caribbean), Europe, Far East (including Australasia), Africa, Middle East (including India and the Indian Ocean), and South America. References are cited as the initials of the first or sole author plus a two-digit date (20th century assumed), the first initial being that of the family or surname. For example, Bay, T.A. 1977 is abbreviated as bta77. In the case of more than one reference in the Bibliography being represented by the same initials and date, it will usually be evident from the context which reference is intended. A slash indicates more than one author; a hyphen represents a hyphenated name.

Arctic Albian references   None were used.

African Albian references   A total of 35 references was used: as/83, ap/83, bft72, bk74, bgp/66, bpf/83, bh66, bg70, cg/62, fr/79, fr75, fr81, fs/73, gpcde/82, gr75, k-km82, kca76, lz/76, lr76, me73, mpa61, mrc72, psw83, pc75, ru82, rra/76, rd66, sj78, sj80, sj/81, tb/79, wa82, wj82, ww83, wp/82

European Albian references   A total of 80 references was used: ac78, a-va/79, ah64, aj/74, aj/75, ba/78, bgm/69, bpy/79, bpy/78, ba/79, bj76, bph82, cj79b, cp/80, cs76, cm/82, clr74, ci/81, cp/71, cp71, cl/72, cg/79, dra/71, dm/83, dr78, di67a, di67b, di68, dpvo70, dp/78, fp67, fp71, fp/73, frm/75, fw67, g-hm79, gaa/78, gmj63, ga80, ge82, gmp71, hm78, hbb69, jm66, j-sm79, j-sm81, km83, km63, lgk/80, ms/79, mc/80, me74, me79, md/81, mr68, mj/83, ndp81, ogh73, ogh75, ohg79, pb78, pc75, qr69, ra79, rp68, rpf80, rj79, rj/77, rf81, rc73, rn75, sw/69, sb/76, sp/77, tmja71, tka81b, wr67, we73, ww64, zpa82a

Far East Albian references   A total of 22 references was used: aga78, amc79, bdd80, bda/68, bcm/80, dji/76, ejl72, epr81, enf/76, enf/80, hg69, kt74, kt80, lnh78, mt78, os82, ppe75/, sj/80, sgr/78, vjj/82, wgd66, wh/71

Middle East Albian references   A total of 17 references was used: ba76, ba77, br71, bt/80, imw81, imw83, jga/65, kwj/83, lz/76, mrj80, phs79, plf/69, rr/68, sa78, wd68, wjl75, wr/80

North American Albian references   A total of 70 references was used: bta77, bew63, bwl/83, bba70, bwf80, bwr/69, bdj68, cma71, ctp73, cac63, ctd/75, csp83, djc70, d-cz70, drl73, dj79, dj82b, drjw/70, ep74, ep77, ep83, fwl/69, fjm63, gmm77, gar/82, hpt70, jlf/75b, jja71a, jja75, jja76, jja77a, krs77, kkm/71, lm73, l-fj81, lmw/78, l-re81, mcgdp/72, mcgaa/68, mjr/81, maa/74, mad73, mad79, mch/66, ohg73, pbf74, rfg81, rjb57, rdd/80, rra64, sjs81, srw70, srw79, sdf73, sci70, sci81, snf/76, scr75b, sdf68, tm77, wrj83, wr80, wgd/75, whh74, wbj/83, wgr62, ycj/81, yrg60, yfg/76, zpa69.

South American Albian references   A total of 37 references was used: ae/73, bfa74, bp/81, bc83, bve56, bve68, bk70, bh73, cg70, cj79, db78, drh/82, e-sf/76, fa/63, hhj62, i.e.m75, jm73, jh74, jh77, lg78, mdim/83, mn/83, mhg71, mf79, mjc/64, mrc/75, nep/80, oha74, oha82, pe/69, rl67, rra/76, spc/76, sf79, wj82, wjj63, ze/72.

The dominance of references from Europe and North America is clear.

Readers can search additional information on individual Stages and groups of Stages by examining the Bibliography.

_ 5 _
Changes in land area through time

Regression, or the retreat of the sea, ideally produces a characteristic lateral sequence in the stratigraphic record, also seen in a vertical section. Similarly, advances of the sea, or transgressions, produce sequences of a different but equally distinct character.

The most important influence on the position of shorelines worldwide is global sea level. It has been generally recognised in recent years that geologically rapid changes in global sea level are not only a characteristic of Ice Ages, such as the Quaternary, but also occur at a rather less rapid frequency, throughout the geological record (Haq, Hardenbol and Vail 1988). Certainly significant changes in global sea level occurred within many of the Stages represented in this atlas. On the whole the paleocoastlines we have indicated correspond to episodes of maximum marine transgression (maximum global sea level) within the time period of the individual map.

Although we leave interpretation and further application of our maps to the interested reader, we have taken the opportunity of using them to estimate changes in land area (and hence in marine inundation) through the Mesozoic and Cenozoic, as this provides a unique source of information on the sea level. The area of land on all maps has therefore been calculated, directly from the digitized data. The distribution of land area through time is plotted in Figure 2 and listed in Table 3. It is difficult to estimate the uncertainties in these estimates, but the results of the calculations are shown to the nearest million square kilometers. This is about the size of the non-terrestrial areas between Australia and Antarctica on the Santonian to Cenomanian maps (Maps 12–15). It is unclear whether these areas had connections to the world ocean, as assumed in Figure 2 and Table 3, or were inland seas.

Except for the Maastrichtian and Campanian, the gross fluctuations in land area are remarkably similar to the changes in sea level inferred by Haq *et al.* (1988), at least on a time-scale of about 25 Ma or more. Such a relationship is to be expected. On the other hand, our sea level curve and that of Haq *et al.* have been derived from quite different datasets. It is not our intention here to discuss the significance of our data in this and other respects, but a recent interpretation of global sea level changes is given by Gurnis (1993).

Table 3. *Non-marine areas and ages*

Map and stratigraphic age	Age (Ma)	Non-marine area in 10^6 km^2
Map 1. Pliocene	5	153
Map 2. Late Miocene (Messinian–Tortonian)	10	152
Map 3. Middle Miocene (Serravallian–Langhian)	12	150
Map 4. Early Miocene (Burdigalian–Aquitanian)	20	150
Map 5. Oligocene	30	147
Map 6. Late Eocene (Priabonian)	37	140
Map 7. Middle Eocene (Bartonian–Lutetian)	45	138
Map 8. Early Eocene (Ypresian)	53	137
Map 9. Paleocene (Thanetian–Danian)	60	138
Map 10. Maastrichtian	70	109
Map 11. Campanian	80	110
Map 12. Santonian	85	118
Map 13. Coniacian	88	118
Map 14. Turonian	90	121
Map 15. Cenomanian	95	118
Map 16. Albian	105	129
Map 17. Aptian	120	131
Map 18. Barremian–Hauterivian	130	125
Map 19. Valanginian–Berriasian	138	127
Map 20. Tithonian	148	130
Map 21. Kimmeridgian	153	130
Map 22. Oxfordian	155	132
Map 23. Callovian	160	128
Map 24. Bajocian (There is no Bathonian map)	170	138
Map 25. Toarcian (There is no Aalenian map)	180	139
Map 26. Pliensbachian	190	142
Map 27. Sinemurian	200	146
Map 28. Hettangian	205	150
Map 29. Late Triassic (Rhaetian–Norian–Carnian)	220	134
Map 30. Middle Triassic (Ladinian–Anisian)	238	139
Map 31. Early Triassic (= Scythian or Spathian–Nammalian–Griesbachian)	245	135

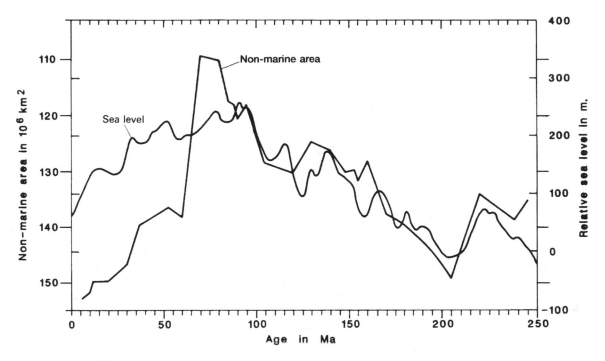

Figure 2. Area of land in millions of square kilometers plotted against time in Ma, superposed on sea level changes in meters versus time in Ma from Haq *et al.* (1988). The graph assumes that the elongate area between Australia and Antarctica for Santonian–Cenomanian time has a connection to the sea. If it was an enclosed sea, the 'non-marine area' for Santonian–Cenomanian time needs to be increased by about 1 million km² for all four stages.

_ 6 _
Relevant literature

The relevant literature listed below comprises references cited in the text and tables together with subsidiary references that have not been specifically cited. The latter include details of Euler rotations and the geographical boundaries of minor continental fragments used in the reconstructions.

Anderson, T.H. and Schmidt, V.A. (1983). The evolution of Middle America and the Gulf of Mexico-Caribbean Sea region during Mesozoic time. *Bull. geol. Soc. Am.* **94**, 941–966.

Berggren, W.A., Kent, D.V., Obradovich, J.D. and Swisher, C.C. (1991). Toward a revised Paleogene geochronology. In: Prothero, D.R. and Berggren, W.A. (eds) *Eocene–Oligocene climatic and biotic evolution*. Princeton University Press, Princeton, 29–45.

Bullard, E., Everett, J.E. and Smith, A.G. (1965). The fit of the continents around the Atlantic. *Phil. Trans. R. Soc. Lond.* **A258**, 41–51.

Cande, S.C. and Kent, D.V. (1992). A new geomagnetic polarity time scale for the Late Cretaceous and Cenozoic. *J. geophys. Res.* **97**, 13917–13952.

Cande, S.C., LaBrecque, J.L. and Haxby, W.F. (1988). Plate kinematics of the South Atlantic, chron C34 to present. *J. geophys. Res.* **93**, 13479–13492.

Cande, S.C. and Leslie, R.B. (1986). Late Cenozoic tectonics of the southern Chile trench. *J. geophys. Res.* **91**, 471–496.

Cande, S.C. and Mutter, J.C. (1982). A revised identification of the oldest sea-floor spreading anomalies between Australia and Antarctica. *Earth Planet. Sci. Lett.* **58**, 151–160.

Cochran, J.R. (1981). The Gulf of Aden: structure and evolution of a young ocean basin and continental margin. *J. geophys. Res.* **86**, 263–287.

Cox, A.V. and Hart, R.B. (1986). *Plate tectonics: how it works*. Blackwell Scientific Publications, Oxford, 392pp.

Crook, K.A.W. and Belbin, L. (1978). The southwest Pacific during the last 90 million years. *J. geol. Soc. Aust.* **25**, 23-40.

Daly, M.C., Cooper, M.A., Wilson, I., Smith, D.G. and Hooper, B.G.D. (1991). Cenozoic plate tectonics and basin evolution in Indonesia. *Mar. Petrol. Geol.* **8**, 2–21.

Dawes, P.R. and Kerr, J.W. (1982). The case against major displacement along Nares Strait. In: Dawes, P.R. and Kerr, J.W. (eds.) *Nares Strait and the drift of Greenland: a conflict in plate tectonics. Meddel. Grønland, Geosci.*, **8**, 369–386.

Enkin, R.J., Yang, Z., Chen, Y. and Courtillot, V. (1992). Paleomagnetic constraints on the geodynamic history of the major blocks of China from the Permian to the present. *J. geophys. Res.* **97**, 13953–13990.

Fisher, R.L. and Sclater, J.G. (1983). Tectonic evolution of the southwest Indian Ocean since the mid-Cretaceous: plate motions and stability of the pole of Antarctica/Africa for at least 80 myr. *Geophys. J. R. astron. Soc.* **73**, 553–576.

Francheteau, J. (1973). Plate tectonic model of the opening of the Atlantic Ocean south of the Azores. In: Tarling, D.H. and Runcorn, S.K. (eds.) *Implications of continental drift to the earth sciences,* vol. 1. Academic Press, London & New York, 197–202.

Funnell, B.M. (1990). Global and European shorelines, Stage by Stage. In: Ginsburg, R.N. and Beaudoin, B. (eds.) *Cretaceous resources, events and rhythms*. Kluwer Academic Publishers, Netherlands, 221–235.

Gradstein, F.M., Agterberg, F.P., Ogg, J.G., Hardenbol, J. and Huang, Z. (1993). A Mesozoic time scale. *J. geophys. Res.*, in press.

Gurnis, M. (1993). Phanerozoic marine inundation of continents driven by dynamic topography above subducting slabs. *Nature*, **364**, 589–593.

Haq, B.U., Hardenbol, J. and Vail, P.R. (1988). Mesozoic and Cenozoic chronostratigraphy and eustatic cycles. In: Wilgus, C.K., Hastings, B.S., Posamentier, H., Wagoner, J.V., Ross, C.A. and Kendall, C.G.StC. (eds.) *Sea-level changes: an integrated approach*. Special Publication 42, Society of Economic Paleontologists and Mineralogists, Tulsa, Oklahoma, 71–108.

Harland, W.B., Armstrong, R.L., Cox, A.V., Craig, L.A., Smith, A.G. and Smith, D.G. (1990). *A geologic time scale 1989*. Cambridge University Press, Cambridge, pp. 263.

Klitgord, K.M. and Schouten, H. (1990). Plate kinematics of the central Atlantic. In: Vogt, P.R. and Tucholke, B.E. (eds.) *Geology of North America*, vol. M, *The western North Atlantic region*. Geological Society of America, 351–378.

Lawver, L.A., Sclater, J.G. and Meinke, L. (1985). Mesozoic and Cenozoic reconstructions of the South Atlantic. *Tectonophysics* **114**, 233–254.

LeFort, J.-P. and Van der Voo, R. (1981). A kinematic model for the collision and complete suturing between Gondwanaland and Laurussia in the Carboniferous. *J. Geol.* **89**, 537–550.

LePichon, X. and Francheteau, J. (1978). A plate-tectonic analysis of the Red Sea – Gulf of Aden area. *Tectonophysics* **46**, 369–406.

Lock, J. and McElhinny, M.W. (1991). The global paleomagnetic database: design, installation and use with ORACLE. *Surv. Geophys.* **12**, 317–491.

McKenzie, D.P., Davies, D. and Molnar, P. (1970). Plate tectonics of the Red Sea and East Africa. *Nature* **226**, 243–248.

Minster, J.B. and Jordan, T.H. (1978). Present-day plate motions. *J. geophys. Res.* **83**, 5331–5354.

Mitchell, A.H.G. (1981). Phanerozoic plate boundaries in mainland SE Asia, the Himalayas and Tibet. *J. geol. Soc. Lond.* **138**, 109–122.

Mutter, J.C., Hegarty, K.A., Cande, S.C. and Weissel, J.K. (1985). Breakup between Australia and Antarctica: a brief review in the light of new data. *Tectonophysics* **114**, 255–279.

Norton, I.O. and Sclater, J.G. (1979). A model for the evolution of the Indian Ocean and the breakup of Gondwanaland. *J. geophys. Res.* **84**, 6803–6830.

Nunns, A.G. (1982). Plate tectonic evolution of the Greenland–Scotland ridge and surrounding regions. In: Bott, M.H.P., Saxov, S., Talwani, H. and Thiede, J. (eds.) *Structure and development of the Greenland–Scotland ridge: new methods and concepts*, vol. 4. NATO Conference Series, IV Marine Series, IV. Plenum, New York, 11-30.

Olivet, J.L. (1978). Nouveau modèle d'évolution de l'Atlantique Nord et Central. Ph.D. thesis, Paris 230 pp.

Piper, J.D.A. (1988). *Palaeomagnetic Database*. Open University Press, Milton Keynes, 264 pp.

Rabinowitz, P.D. and LaBrecque, J. (1979). The Mesozoic South Atlantic Ocean and evolution of its continental margins. *J. geophys. Res.* **84**, 5973–6002.

Rehault, J.-P., Boillot, G. and Mauffret, A. (1985): The western Mediterranean basin. In: Stanley, D.J. and Wezel, F.-C. (eds.) *Geological evolution of the Mediterranean Basin*. Springer-Verlag, New York, 101–129.

Roest, W.R. and Srivastava, S.P. (1989). Sea floor spreading in the Labrador Sea: a new reconstruction. *Geology* **17**, 1000–1003.

Sclater, J.G., Hellinger, S. and Tapscott, C. (1977). The palaeobathymetry of the Atlantic Ocean from the Jurassic to the present. *J. Geol.* **85**, 509–552.

Sclater, J.G., Fisher, R.L., Patriat, P., Tapscott, C. and Parsons, B. (1981). Eocene to recent development of the South-west Indian ridge, a consequence of the evolution of the Indian Ocean triple junction. *Geophys. J. R. astron. Soc.* **64**, 587–604.

Segoufin, J. and Patriat, P. (1981). Reconstructions of the western Indian Ocean at anomalies M21, M2 and 34 times. Palaeoposition of Madagascar. *Bull. Soc. géol. France* **23**, 603–607.

Sengor, A.M.C., Yilmaz, K. and Ketin, I. (1980). Remnants of a pre-Late Jurassic ocean in northern Turkey: fragments of a Permian-Triassic paleo-Tethys? *Bull. geol. Soc. Am.* **91**, 599–609.

Sibuet, J.-C., Ryan, W.B.F., Arthur, M., Barnes, R., Blechsmidt, G., De Charpel, O., De Graciansky, P.L., Habib, D., Iaclarino, S., Johnson, D., Lopatin, B.G., Maldonado, A., Montadert, L., Moore, D.G., Morgan, G.E., Mountain, G., Rehault, J.P., Sigal, J. and Williams, C.A. (1980). Deep drilling results of Leg 47b (Galicia Bank area) in the framework of the early evolution of the North Atlantic Ocean. *Phil. Trans. R. Soc. Lond. A* **294**, 51–61.

Smith, A.G. and Hallam, A. (1970). The fit of the southern continents. *Nature* **225**, 139–144.

Smith, A.G. and Livermore, R.A. (1991). Pangea in Permian to Jurassic time. *Tectonophysics* **187**, 135–179.

Smith, A.G., Briden, J.C. and Drewry, G.E. (1973). Phanerozoic world maps. In: Hughes, N.F. (ed.) *Organisms and continents through time.* Special Papers on Palaeontology, London, no.**12**, 1–42.

Smith, A.G., Hurley, A.M. and Briden, J.C. (1981). *Phanerozoic palaeocontinental world maps.* Cambridge University Press, Cambridge, 102 pp.

Smith, D.G. (1987). Late Paleozoic to Cenozoic reconstructions of the Arctic. In: Tailleur, I. & Weimer, P. (eds.) *Alaskan North Slope geology.* SEPM Special Publication, 785–795.

Srivastava, S.P. and Tapscott, C.R. (1986). Plate kinematics of the North Atlantic. In: Vogt, P.R. and Tucholke, B.E. (eds.) *Geology of North America,* vol. M, *The western North Atlantic region.* Geol. Soc. America, 379–404.

Stock, J. and Molnar, P. (1982). Uncertainties in the relative positions of the Australia, Antarctica, Lord Howe, and Pacific plates since the late Cretaceous. *J. geophys. Res.* **87**, 4697–4714.

Tyson, R.V. and Funnell, B.M. (1987). European Cretaceous shorelines, stage by stage. *Paleogeog., Paleoclimatol., Paleoecol.* **59**, 69–91.

Van der Voo, R. and French, R.B. (1974). Apparent polar wandering for the Atlantic bordering continents: Late Carboniferous to Eocene. *Earth Sci. Rev.* **10**, 99–119.

Van den Berg, J. and Zijderveld, J.D.A. (1982). Palaeomagnetism in the Mediterranean area. In: Berckhemer, H. and Hsu, K. (eds.) *Alpine Mediterranean geodynamics* vol. 7. Geodynamics Series, 7. American Geophysical Union, Washington, DC, 83–112.

Vogt, P.R., Taylor, P.T., Kovaks, L.C. and Johnston, G.L. (1982). The Canada Basin: aeromagnetic constraints on structure and evolution. In: Johnson, J.L and Sweeney, J.F. (eds.) *Structure of the Arctic. Tectonophysics* **89**, 295–336.

Walper, J.L. (1980). Tectonic evolution of the Gulf of Mexico. In: Pilger, R.H. (ed.) Origin of the Gulf of Mexico and the early opening of the central North Atlantic Ocean. Louisiana State University, Baton Rouge, 87–98.

Weissel, J.M. and Hayes, D.E. (1972). Magnetic anomalies in the southeast Indian Ocean. In: Hayes, D.E. (ed.) *Antarctic Oceanology II: The Australian–New Zealand sector,* vol. 19. Antarctic Research Series, **19**, 165–196.

Weissel, J.K. and Hayes, D.E. (1977). Evolution of the Tasman Sea reappraised. *Earth Planet. Sci. Lett.* **36**, 77–84.

Weissel, J.K., Hayes, D.E. and Herron, E.M. (1977). Plate tectonic synthesis: the displacements between Australia, New Zealand and Antarctica since the Late Cretaceous. *Mar. Geol.* **25**, 231–277.

Ziegler, A.M., Scotese, C.R. and Barrett, S.F. (1982). Mesozoic and Cenozoic paleogeographic maps. In: Brosche and Sundermann, P.J. (eds.) *Tidal friction and the earth's rotation II,* Springer-Verlag, Berlin, 240–252.

_ 7 _
Paleocoastline maps

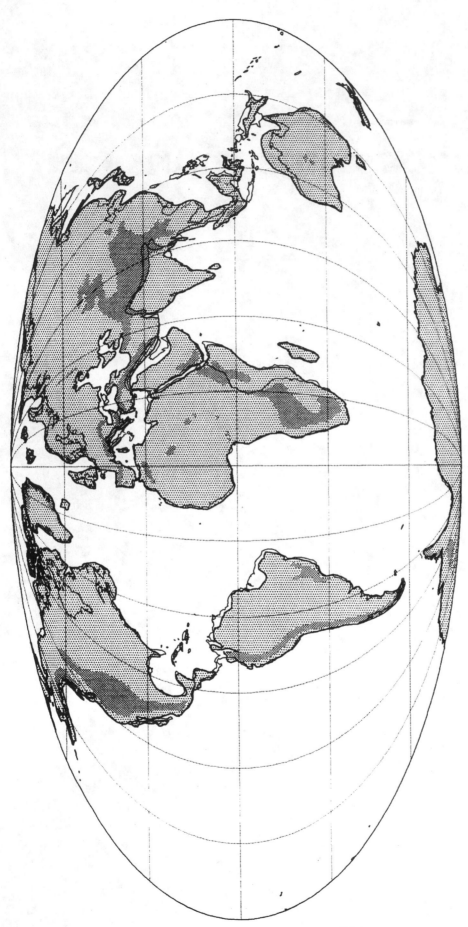

Map 1. Pliocene 5 Ma paleocoastline map.

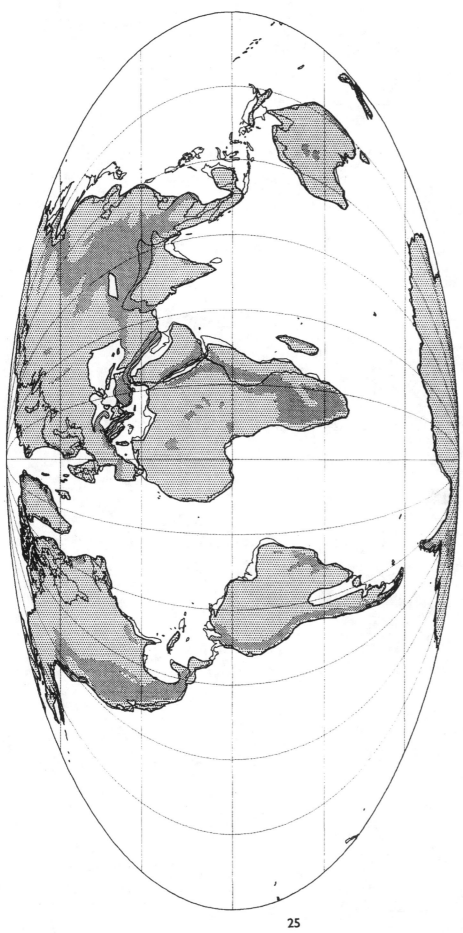

Map 2. Late Miocene (Messinian–Tortonian) 10 Ma paleocoastline map.

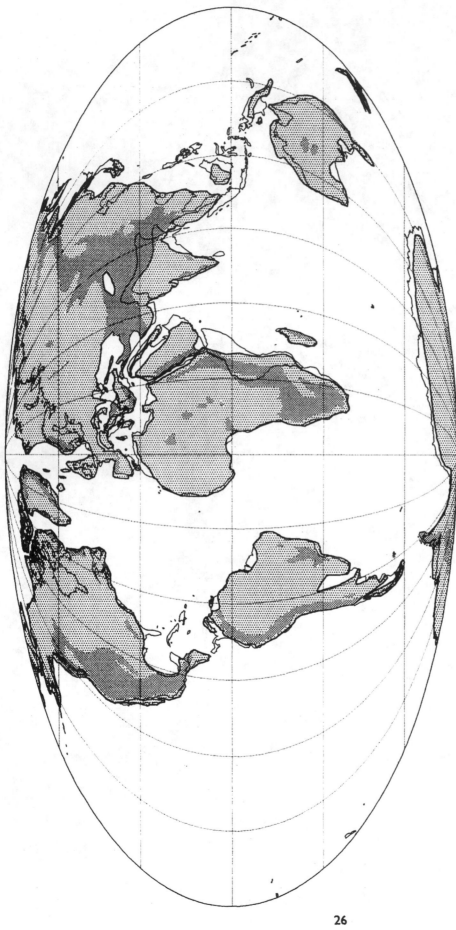

Map 3. Middle Miocene (Serravallian–Langhian) 12 Ma paleocoastline map.

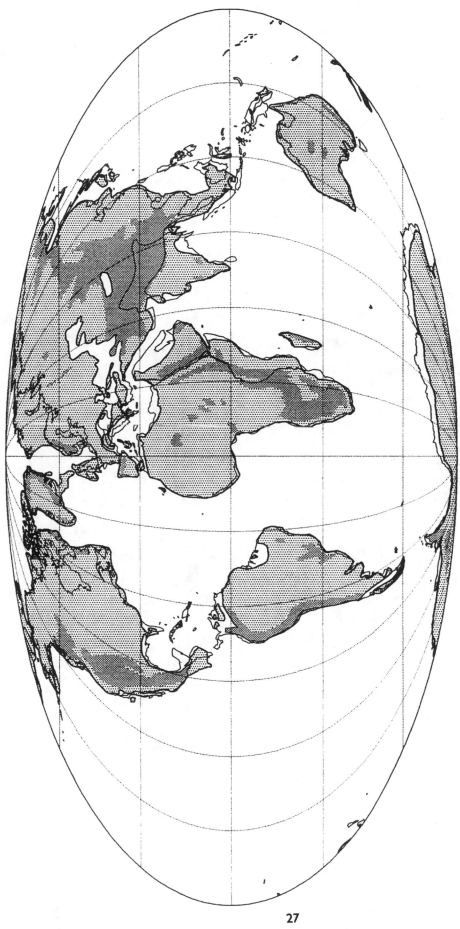

Map 4. Early Miocene (Burdigalian–Aquitanian) 20 Ma paleocoastline map.

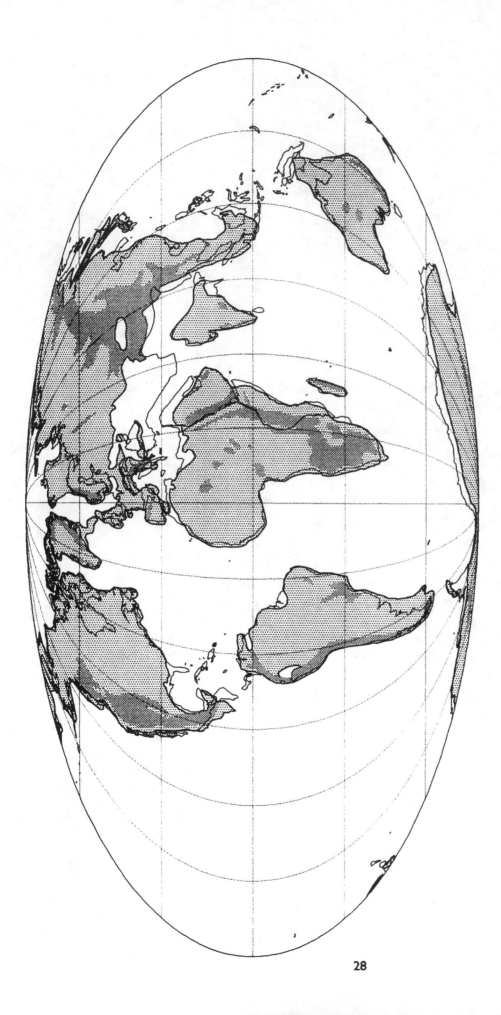

Map 5. Oligocene 30 Ma paleocoastline map.

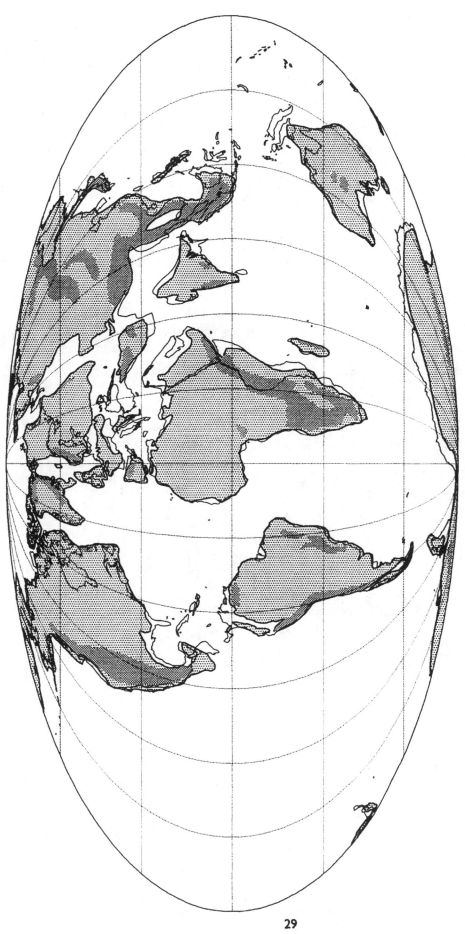

Map 6. Late Eocene (Priabonian) 37 Ma paleocoastline map.

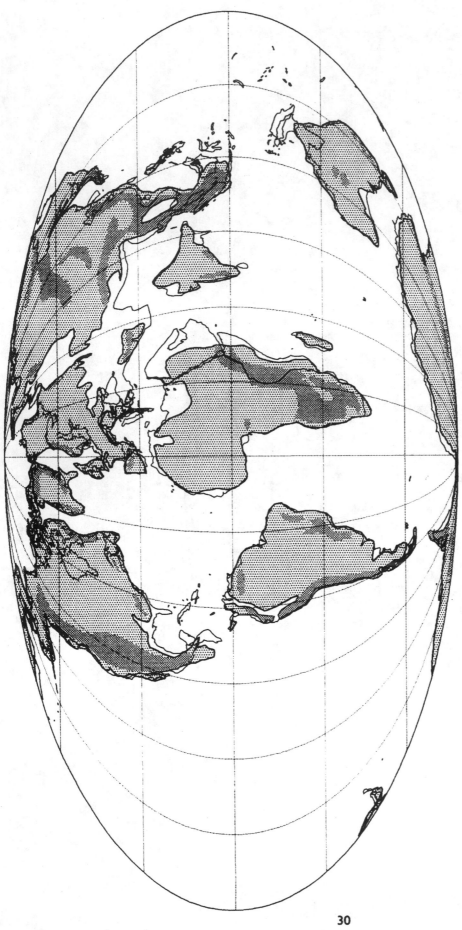

Map 7. Middle Eocene (Bartonian–Lutetian) 45 Ma paleocoastline map.

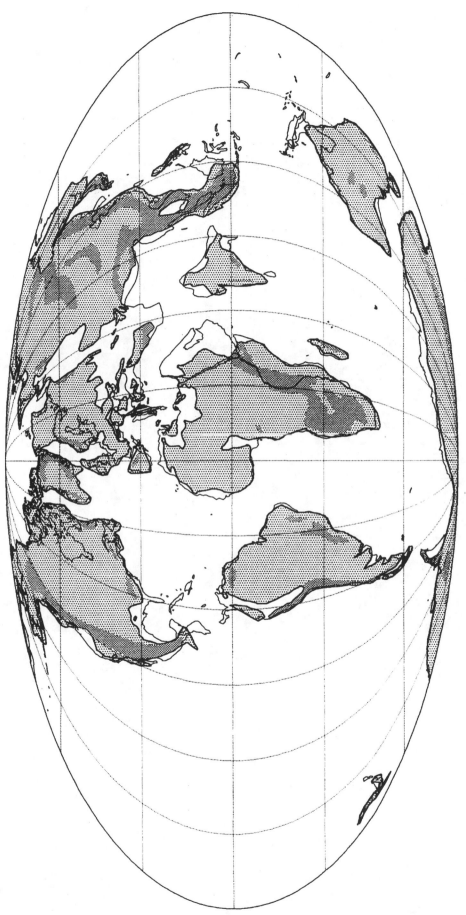

Map 8. Early Eocene (Ypresian) 53 Ma paleocoastline map.

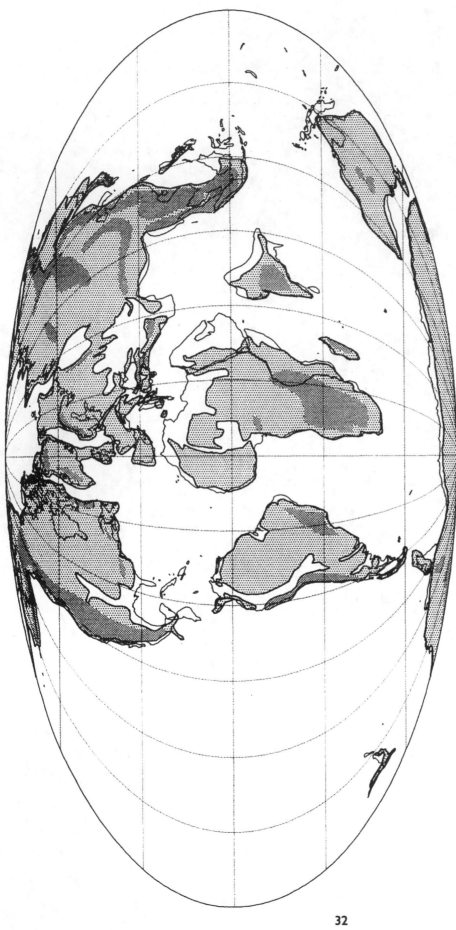

Map 9. Paleocene (Thanetian–Danian) 60 Ma paleocoastline map.

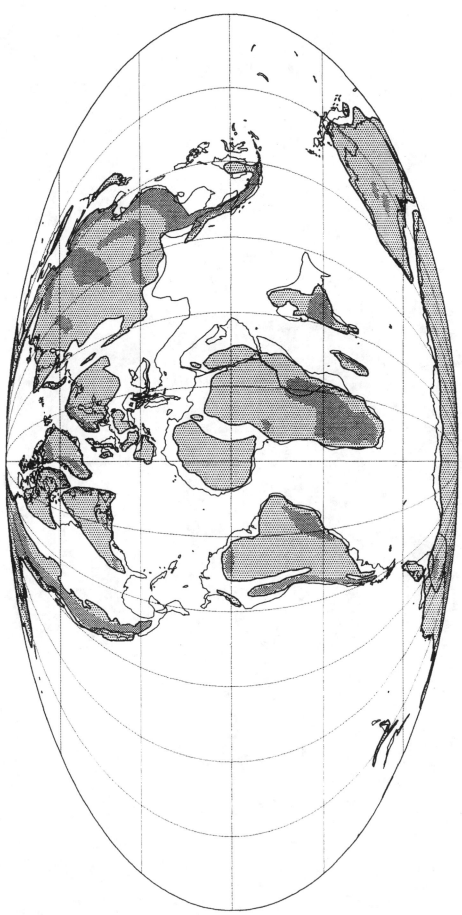

Map 10. Maastrichtian 70 Ma paleocoastline map.

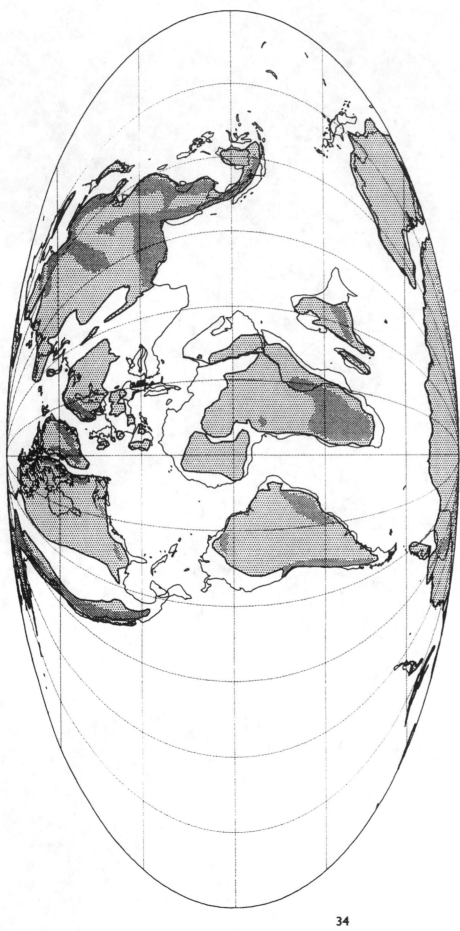

Map 11. Campanian 80 Ma paleocoastline map.

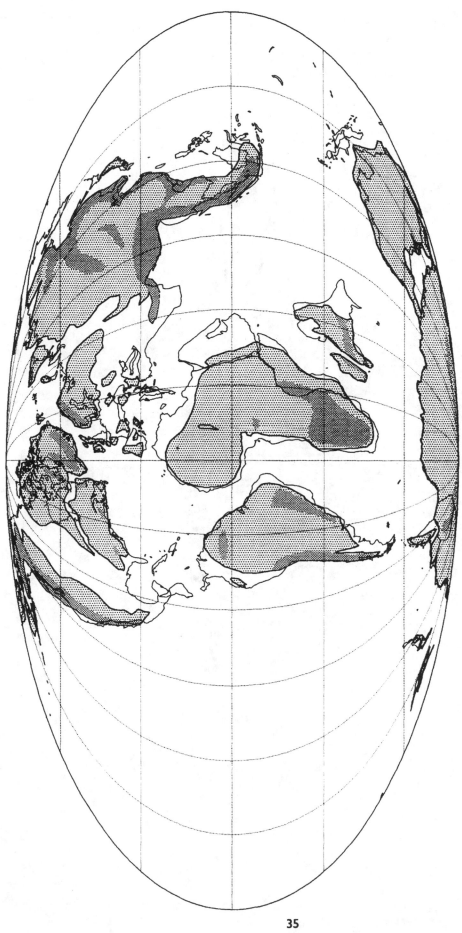

Map 12. Santonian 85 Ma paleocoastline map.

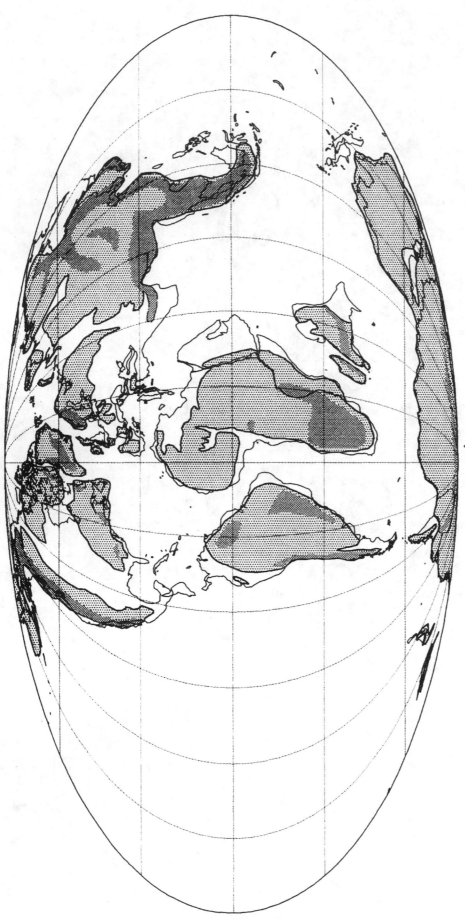

Map 13. Coniacian 88 Ma paleocoastline map.

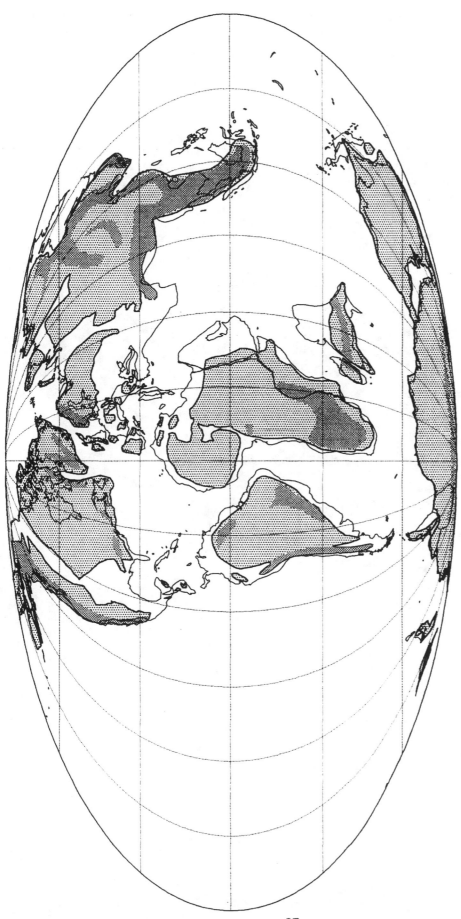

Map 14. Turonian 90 Ma paleocoastline map.

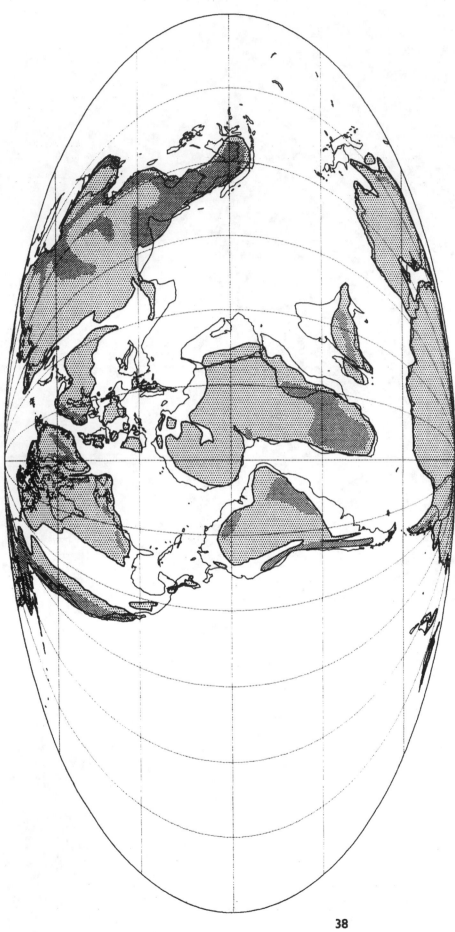

Map 15. Cenomanian 95 Ma paleocoastline map.

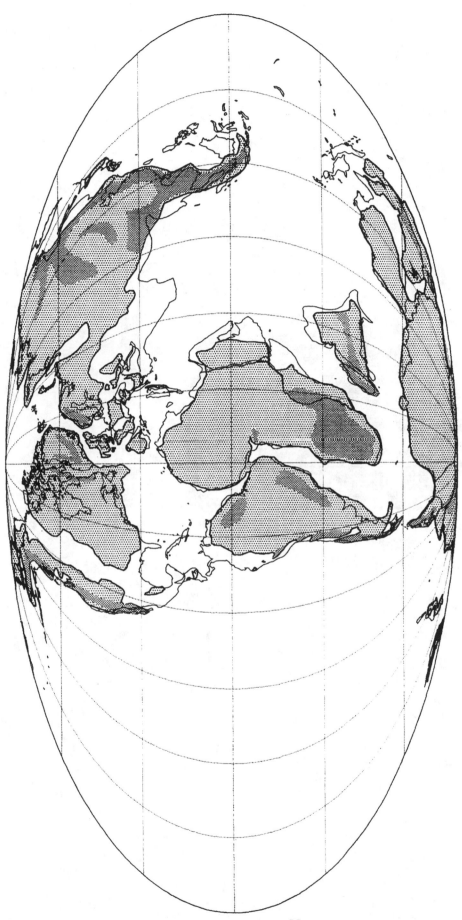

Map 16. Albian 105 Ma paleocoastline map.

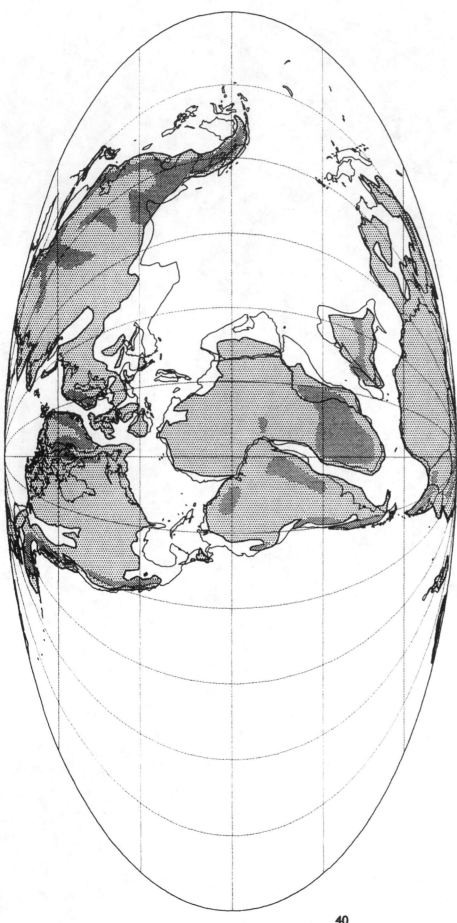

Map 17. Aptian 120 Ma paleocoastline map.

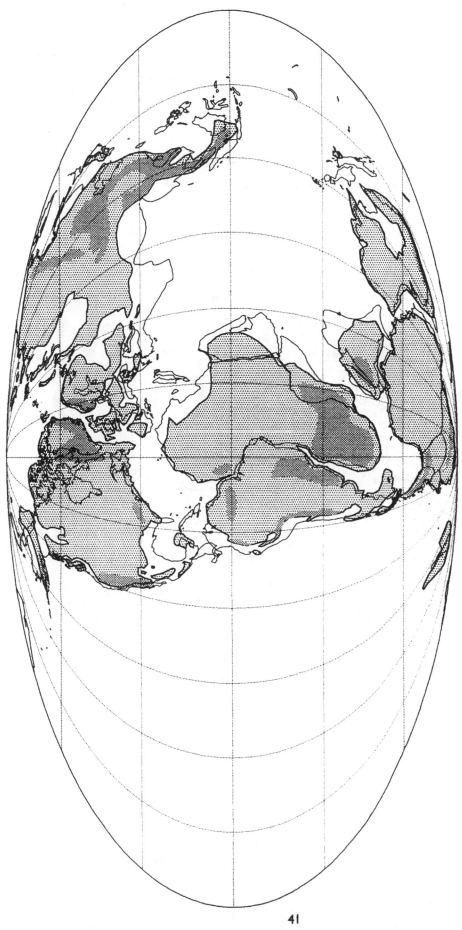

Map 18. Barremian–Hauterivian 130 Ma paleocoastline map.

41

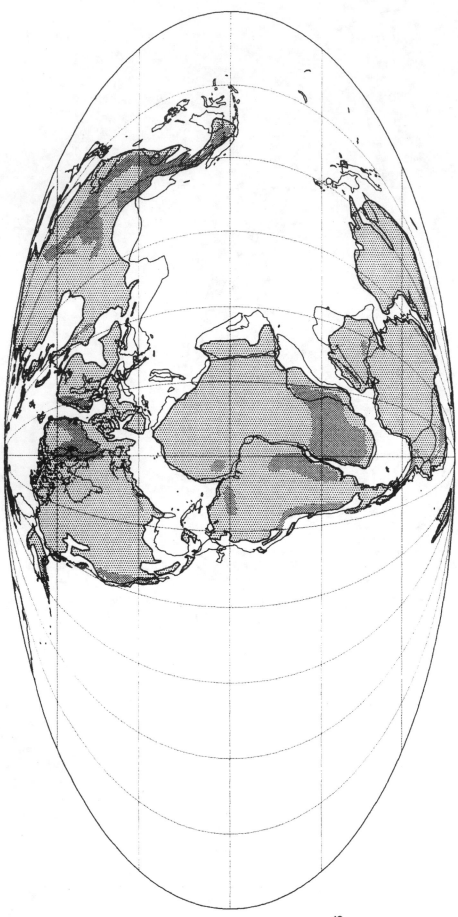

Map 19. Valanginian–Berriasian 138 Ma paleocoastline map.

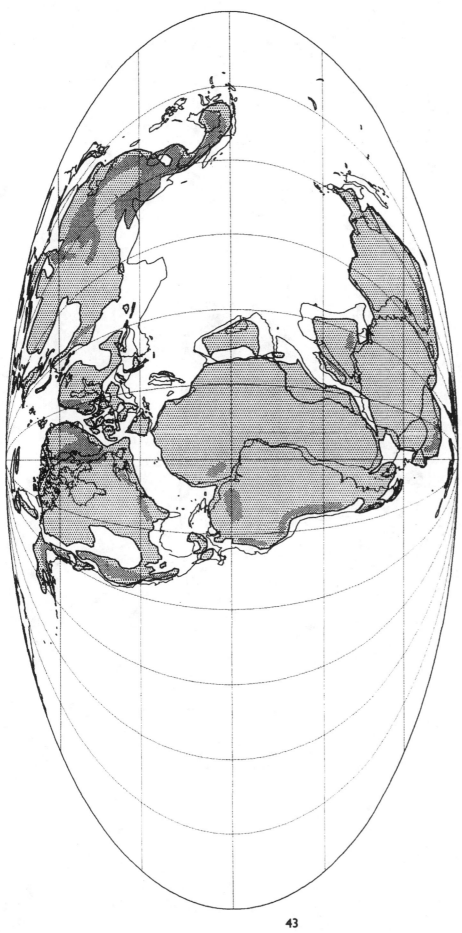

Map 20. Tithonian 148 Ma paleocoastline map.

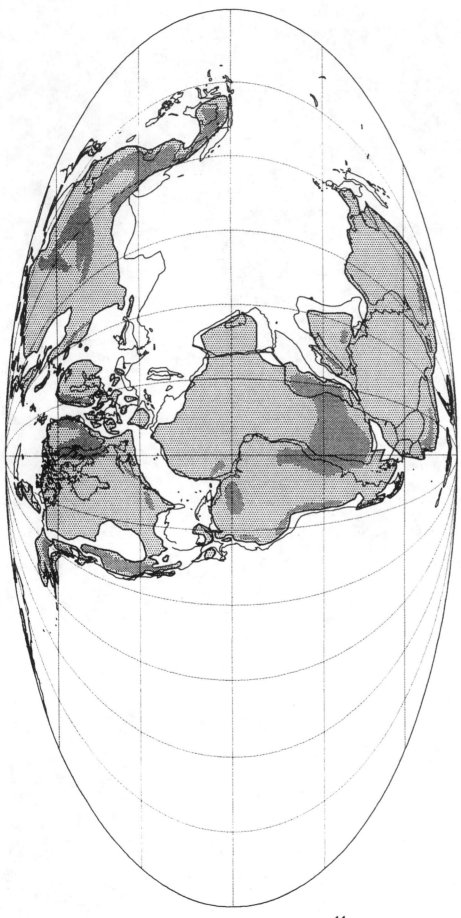

Map 21. Kimmeridgian 153 Ma paleocoastline map.

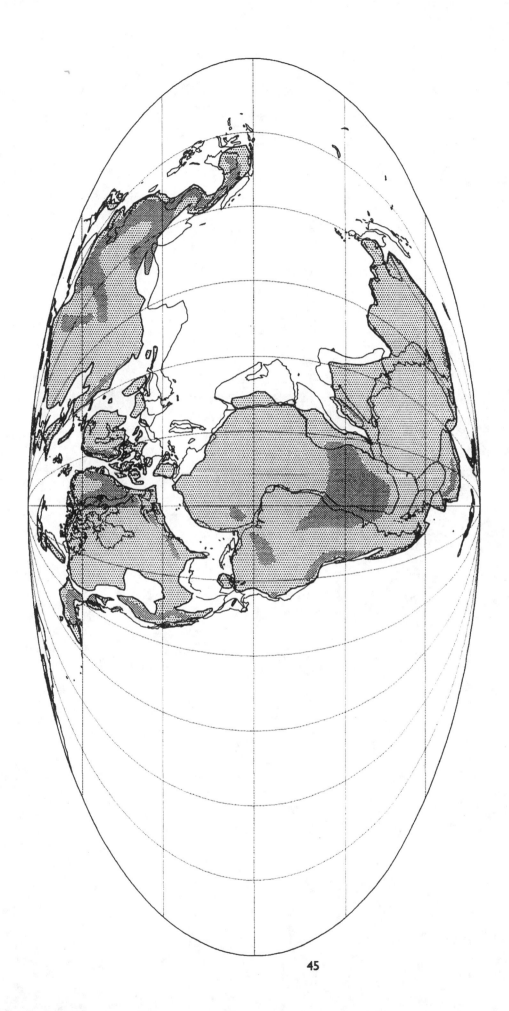

Map 22. Oxfordian 155 Ma paleocoastline map.

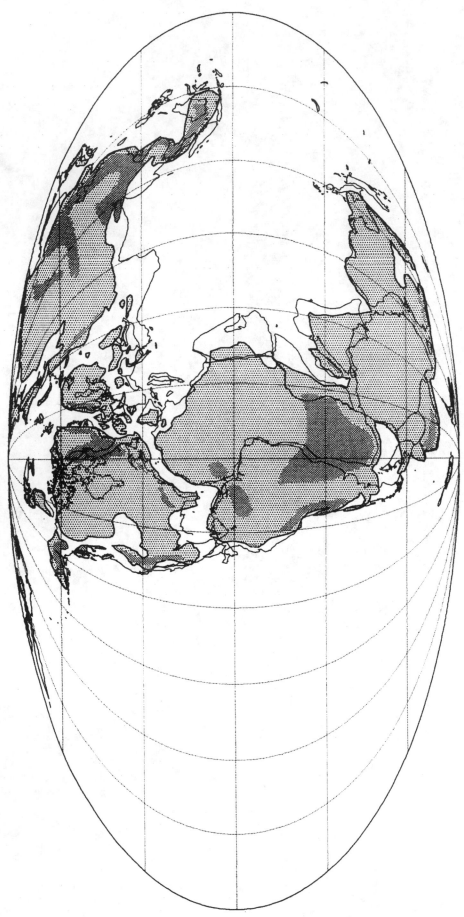

Map 23. Callovian 160 Ma paleocoastline map.

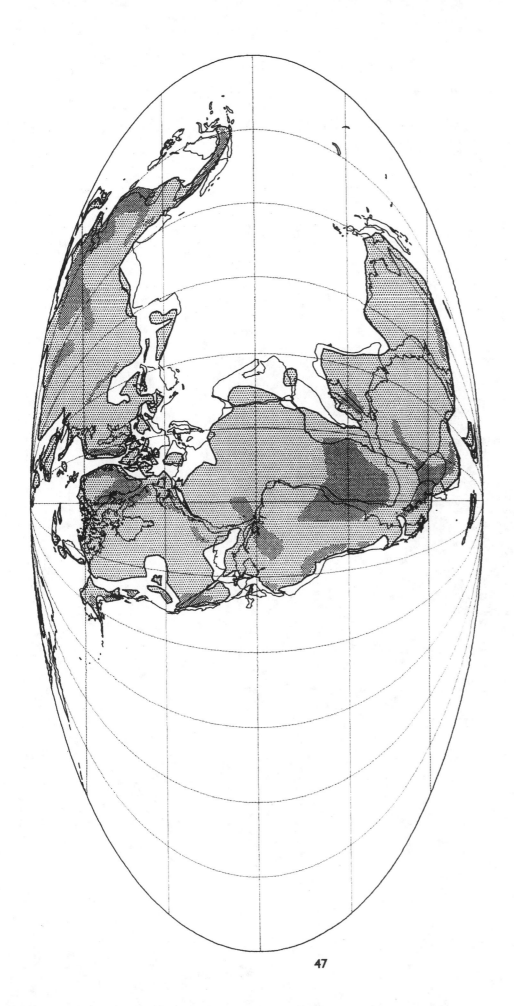

Map 24. Bajocian 170 Ma paleocoastline map. There is no Bathonian map.

47

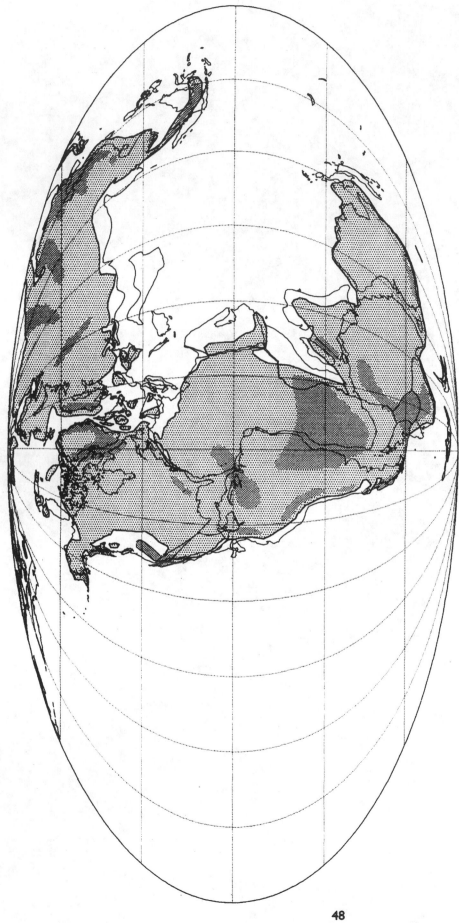

Map 25. Toarcian 180 Ma paleocoastline map. There is no Aalenian map.

48

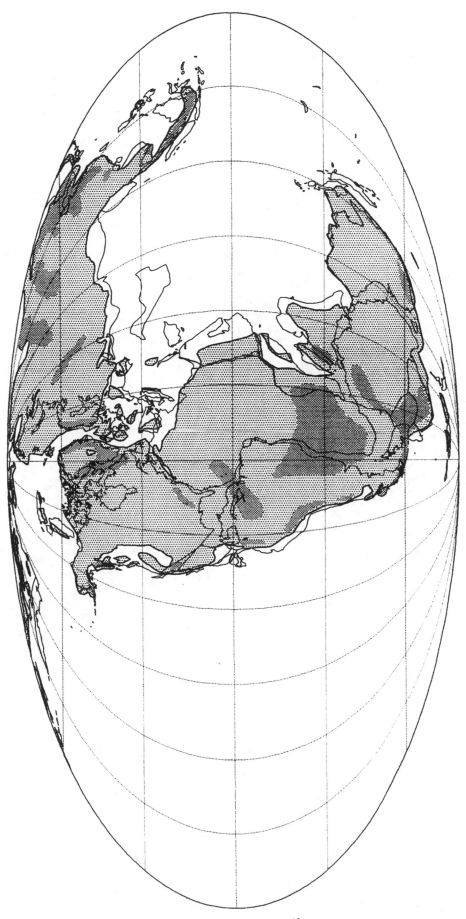

Map 26. Pliensbachian 190 Ma paleocoastline map.

49

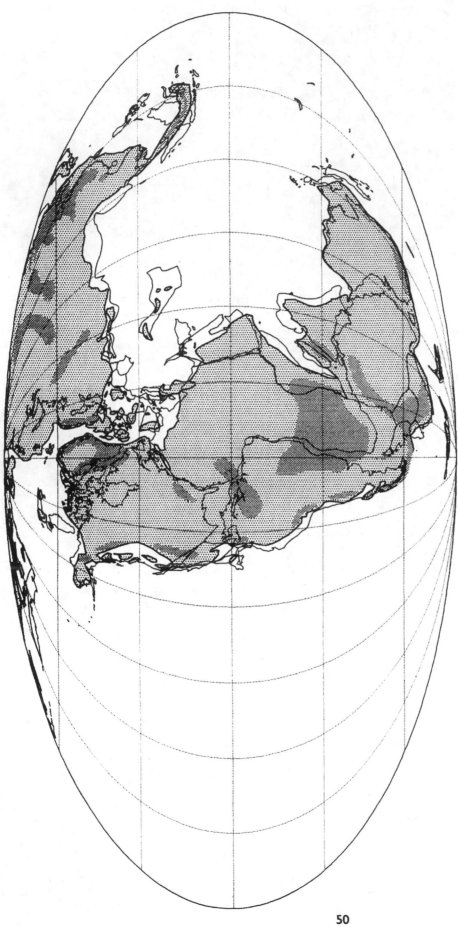

Map 27. Sinemurian 200 Ma paleocoastline map.

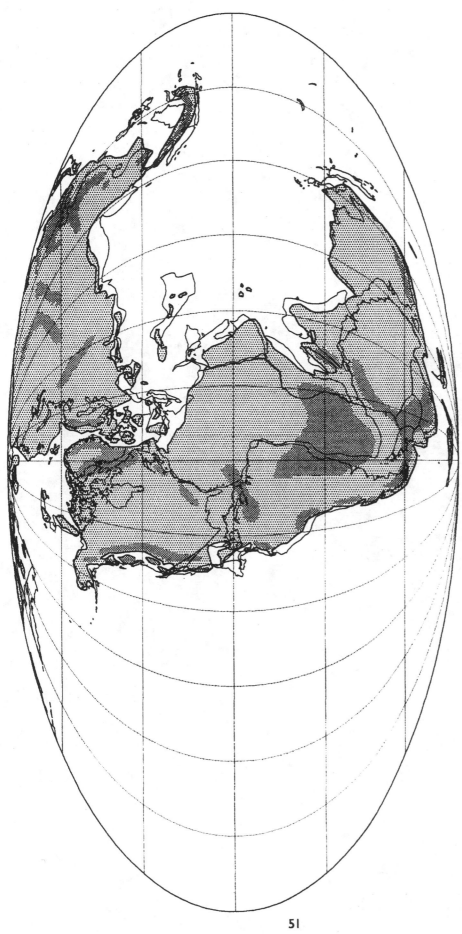

Map 28. Hettangian 205 Ma paleocoastline map.

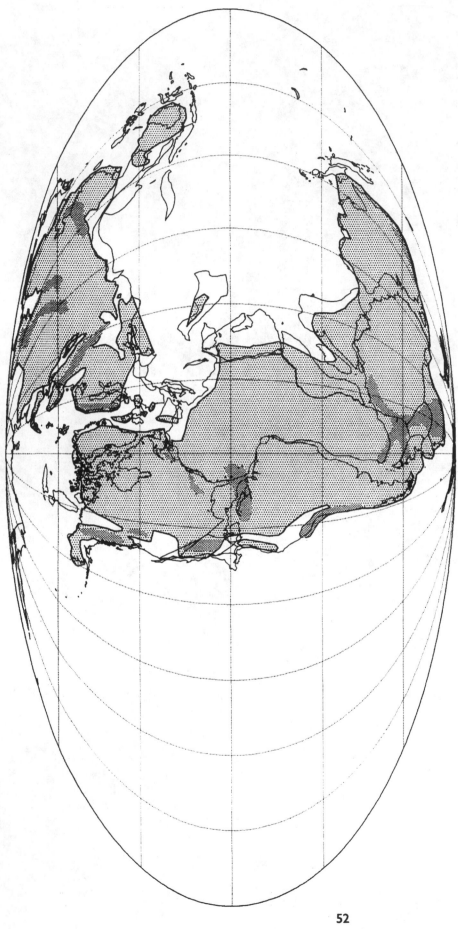

Map 29. Late Triassic (Rhaetian–Norian–Carnian) 220 Ma paleocoastline map.

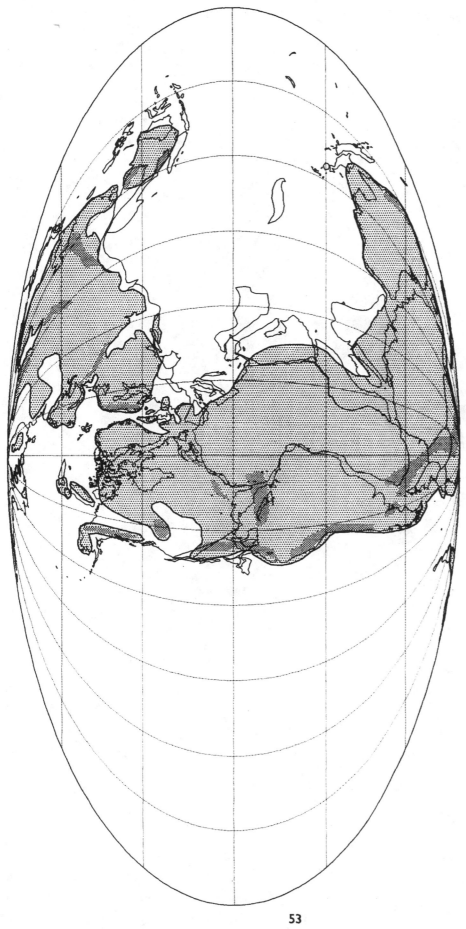

Map 30. Middle Triassic (Ladinian–Anisian) 238 Ma paleocoastline map.

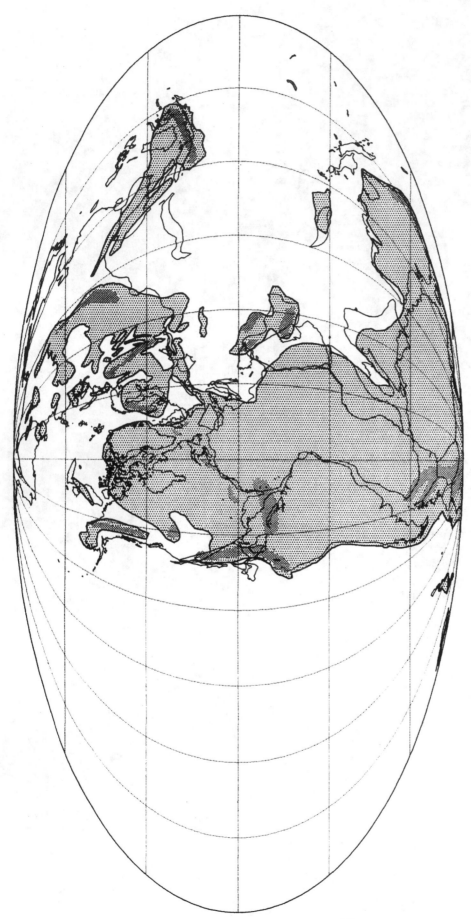

Map 31. Early Triassic (= Scythian or Spathian–Nammalian–Griesbachian) 245 Ma paleocoastline map.

– 8 –
Bibliography

Abbate, E. and Sagri, M. (1970). The eugeosynclinal sequence. In: Sestini, G. (ed.) *Development of the northern Apennines geosyncline. Sediment. Geol.* **4** (3–4), 251–340.

Abdallah, A.M., El Addani, A. and Fahmy, N. (1963). Stratigraphy of the Lower Mesozoic rocks western side of Gulf of Suez, Egypt. *Geol. Surv. min. Res. Dep. Pap. Cairo.* **27**, 23pp.

Abdel-Monem, A.A. and Kulp, J.L. (1968). Paleogeography and the source of sediments of the Triassic basin, New Jersey, by K-Ar dating. *Bull. geol. Soc. Am.* **79** (9), 1231–1242.

Abjean, M. (1980). Premiers éléments de l'étude d'une plate-forme carbonatée: Le Portlandien Berriasien du Verdon (Var-Alpes de Haute Provence). In: Philip, J. (ed.) *Journées de la R.C.P. 510, Univ. Provence, Marseille Nov. 1979. Trav. Lab. Géol. hist. Paléontol. Univ. Provence.* **11**, 21–33.

Adams, A.E. (1979). Sedimentary environments and palaeogeography of the western High Atlas, Morocco, during the Middle and Late Jurassic. *Palaeogeogr. Palaeoclimatol. Palaeoecol.* **28** (3–4), 185–196.

Adeleye, D.R. (1973). Origin of ironstones, an example from the Middle Niger valley, Nigeria. *J. sediment. Petrol.* **43** (3), 709–727.

Adeleye, D.R. (1975). Nigerian late Cretaceous stratigraphy and paleogeography. *Bull. Am. Assoc. Petrol. Geol.* **59** (12), 2302–2313.

Aepler, R. (1974). The Rhaetic sandstone of the Tübingen area – a condensed delta. *Neues Jahrb. Geol. Palaontol. Abhandlungen.* **147** (2), 113–162.

Ager, D.V. (1975). The Jurassic world ocean (with special reference to the N. Atlantic). In: Finstad, K.G. and Selley, R.C. (eds.) *Proc. Symp. Jurassic northern North Sea, Stavanger, Sept. 1975, Nor. Pet. Foren.*, **JNNSS/1**, 1–43.

Aguayo, J. (1978). Facies sedimentarias y diagénesis de la Formación Novillo (Jurasico Superior) noreste de México. *Rev. Inst. Mex. Pet.* **10** (4), 6–45.

Aharoni, E. (1966). Oil and gas prospects of Kurnub Group (Lower Cretaceous) in southern Israel. *Bull. Am. Assoc. Petrol. Geol.* **50** (11), 2388–2403.

Ahmad, F. (1961). Palaeogeography of the Gondwana period in Gondwanaland, with special reference to India and Australia, and its bearing on the theory of continental drift. *Mem. geol. Surv. India.* **90**, 142pp.

Aiello, E., Bartolini, C., Boccaletti, M., Gocev, P., Karagjuleva, J., Kostadinov, V. and Manetti, P. (1977). Sedimentary features of the Srednogorie zone (Bulgaria): an Upper Cretaceous intra-arc basin. *Sediment. Geol.* **19** (I), 39–68.

Ainardi, R. (1975). Microfacies et paléoécologie du 'Purbeckien' des régions de Nantua et Saint-Rambert (Jura méridional). In: *Proc. Symp., Colloque sur la limite Jurassique-Crétacé, Lyon, Sept. 1973, Mém. Bur. Rech. géol. minières.* **86**, 232–241.

Ainardi, R. (1977a). Un exemple de sédimentation lagunaire et de stabilisation de la ligne de rivage: les niveaux dolomitiques de la Lettenkohle dans les Vosges sud-occidentales. *Bull. Soc. géol. Fr. Sér. 7.* **19** (2), 249–256.

Ainardi, R. (1977b). Un paysage margino-littoral 'le Purbeckien' du Jura méridional. *Bull. Soc. géol. Fr. Sér. 7.* **19** (2), 257–264.

Ala, M.A. and Moss, B.J. (1978-9). Comparative petroleum geology of southeast Turkey and northeast Syria. *J. Petrol. Geol.* **1** (4), 3–27.

Alavi-Naini, M. (1972). Etude géologique de la région de Djam. *Rep. geol. Surv. Iran.* **23**, 288pp.

Albers, H.J. (1976). Feinstratigraphie, Faziesanalyse und Zyklen des Untercampans (Vaalser Grünsand – Hervien) von Aachen und dem niederländisch-belgischen Limburg. *Geol. Jahrb.* **A34**, 3–68.

Aldinger, H. (1968). Die Palaeogeographie des Schwäbischen Jurabeckens. *Eclog. geol. Helv.* **61** (1), 167–182.

Alidou, S. and Lang, J. (1983). Etude sédimentologique, paléogéographique et stratigraphique du bassin intracratonique palaeozoïque-mésozoïque de Kandi (Nord-Est Benin – Afrique de l'ouest). *Geol. Rdsch.* **72** (1), 191–205.

Al-Kasim, F., Ronnevik, H.C. and Ulleberg, K. (1975). Review of the Jurassic offshore Norway. In: Finstad, K.G. and Selley, R.C. (eds.) *Proc. Symp. Jurassic northern North Sea, Stavanger, Sept. 1975, Nor. Pet. Foren.*, **JNNSS/3**, 1–18.

Allen, G.A., Pearce, L.G.G. and Gardner, W.E. (1978). A regional interpretation of the Browse Basin. *Aust. Petrol. Explor. Assoc. J.* **18** (1), 23–33.

Allen, P. (1965). L'âge du Purbecko-Wealdien d'Angleterre. In: *Proc. Colloque sur le Crétacé Inférieur, Lyon, Sept. 1963, Mém. Bur. Rech. géol. minières.* **34**, 321–326.

Allen, P. (1967). Origin of the Hastings facies in north-western Europe. *Proc. Geol. Assoc. Lond.* **78** (1), 27–105.

Allen, P. (1975). Wealden of the Weald: a new model. *Proc. Geol. Assoc. Lond.* **86** (4), 389–437.

Allix, P. and Popoff, M. (1983). Le Crétacé Inférieur de la partie nord-orientale du fosse de la Benoue (Nigeria): un exemple de relation étroite entre tectonique et sédimentation. *Bull. Cent. Rech. Explor. Prod. Elf-Aquitaine.* **7** (1), 349–359.

Al-Maleh, A.K. (1976). *Etude stratigraphique, pétrographique, sédimentologique et géochimique du Crétacé du N.W. Syrien (Kurd Dagh et environs d'Aafrine) – les aspects pétroliers de la région.* Unpubl. Ph.D. Thesis. Univ. Pierre et Marie Curie, Paris. **2**, 351–620.

Al-Maleh, A.K. (1982). Evolution de la sédimentation carbonatée dans le nord-ouest de la Syrie (Kurd Dagh et environs d'Aafrine) pendant le Crétacé. In: Lang, J. (ed.) *Livre Jubilaire Gabriel Lucas: géologie sédimentaire. Mém. Géol. Univ. Dijon.* **7**, 473–479.

Almeras, Y. (1965). Etat des connaissances actuelles sur la répartition stratigraphique des Brachiopodes du Crétacé Inférieur (France et pays limitrophes). In: *Proc. Colloque sur le Crétacé Inférieur, Lyon, Sept.*

1963, Mém. Bur. Rech. géol. minières. **34**, 379–392.

Almeras, Y. and Elmi, S. (1982). Fluctuations des peuplements d'ammonites et de brachiopodes en liaison avec les variations bathymétriques pendant le Jurassique inférieur et moyen en Méditerranée Occidentale. In: *Shallow Tethys, Int. Symp. Univ. Padova June 1981 (1). Boll. Soc. geol. Palaeontol. Ital.* **21** (2–3), 196-288.

Al-Omari, F.S. and Sadek, A. (1974). New contributions to the Upper Senonian stratigraphy in northern Iraq. *Geol. Rdsch.* **63** (3), 1217–1231.

Alonso, A. and Floquet, M. (1982). Sédimentation et environnement au Turonien en Vieille Castille (Espagne): un modèle d'évolution en domaine de plate-forme. In: *Proc. Colloque sur le Turonien, Paris, Oct. 1981, Mém. Mus. Nat. Hist. nat. Paris, sér. C.* **49**, 113–128.

Althen, G.W., Rusbult, J. and Seeger, J. (1980). Regionale Neubearbeitung des Muschelkalks. In: Lauterbach, R. (ed.) *Trias Kolloquium 1977, Eisenach, D.D.R., 1977.* **8** (8), 985–999.

Alvarado, M. (1980). Espagne In: Géologie des pays européens: Espagne, Grèce, Italie, Portugal, Yougoslavie. Com. Nat. Fr. géol., 26th Int. geol. congr., Paris, 1980. 1–54.

Andel, Tj.H. van, Thiede, J., Sclater, J.G. and Hay, W.W. (1977). Depositional history of the South Atlantic ocean during the last 125 million years. *J. Geol.* **85** (6), 651–698.

Anderson, R.Y. and Kirkland, D.W. (1960). Origin, varves and cycles of Jurassic Todilto Formation, New Mexico. *Bull. Am. Assoc. Petrol. Geol.* **44** (1), 37–52.

Anderson, T.H. and Schmidt, V.A. (1983). The evolution of Middle America and the Gulf of Mexico-Caribbean Sea region during Mesozoic time. *Bull. geol. Soc. Am.* **94** (8), 941–966.

Antunes, M.T., Ferreira, M.P., Rocha, R.B., Soares, A.F. and Zbyszewski, G. (1980). Portugal: 3. Essai de reconstitution paléogéographique par cycles orogéniques: 3.2 Le cycle alpin. In: *Géologie des pays européens: Espagne, Grèce, Italie, Portugal, Yougoslavie. Com. Nat. Fr. géol., 26th Int. geol. congr., Paris, 1980.* 103–149.

Apthorpe, M.C. (1979). Depositional history of the Upper Cretaceous of the Northwest Shelf, based upon foraminifera. *Aust. Petrol. Explor. Assoc. J.* **19** (1), 74–89.

Arias, C. (1978). Estratigrafía y paleogeografía del Jurásico Superior y Cretácico Inferior del noreste de la Provincia de Albacete. *Seminarios estratigrafía, ser. monograf., Madrid.* **3**, 299pp.

Arias, C., Mas, R., Garcia, A., Alonso, A., Vilas, L., Rincón, R. and Melédez, N. (1979). Les facies urgoniens et leurs variations pendant la transgression Aptienne dans la partie occidentale de la chaine Ibérique. In: *L'Urgonien des Pays Méditerranéens. Géobios. Spéc. Mém.* **3**, 11–24.

Arnaud, H. (1979). Surfaces d'ablation sous-marines et sédiments barremo-bedouliens remanies par gravité du Barremien au Cénomanien entre le Vercors et le Devoluy (SE de la France). *Géol. Alpine.* **55**, 5–21.

Arnaud, H. (1981). De la plate-forme Urgonienne au Bassin Vocontien: Le Barremo-Bedoulien des Alpes Occidentales entre Isère et Buech (Vercors méridional, Diois oriental et Devoluy). *Mém. Géol. Alpine.* **12** (2), 493pp.

Arnaud, M. and Monleau, C. (1979). Evolution séquentielle et palaeogéographie du Jurassique provencal (de l'Hettangien à l'Oxfordien). In: *La sédimentation Jurassique W. européen. Proc. Symp., Paris, May 1977, Assoc. Sédimentol. Fr. Spéc. Publ.* **1**, 181–192.

Arnaud-Vanneau, A. (1979). Répartition de la microfaune benthique

dans les differents paléomilieux urgoniens. In: L'Urgonien des Pays Méditerranéens. *Géobios. Spéc. Mém.* **3**, 255–276.

Arnaud-Vanneau, A. (1980). Micropaléontologie, paléoécologie et sédimentologie d'une plate-forme carbonatée de la marge passive de la Téthys: l'Urgonien du Vercors septentrional et de la Chartreuse (Alpes occidentales). *Mém. Géol. Alpine.* **11** (1), 267pp.

Arnaud-Vanneau, A. and Arnaud, H. (1976). L'évolution paléogéographique du Vercors au Barremien et à l'Aptien inférieur (Chaînes subalpines septentrionales, France). *Géol. Alpine.* **52**, 5–30.

Arnaud-Vanneau, A., Arnaud H., Cotillon, P., Ferry, S. and Masse, J-P. (1982). Caractères et évolution des plates-formes carbonatées Périvocontiennes au Crétacé Inférieur, (France sud-est). In: Cotillon, P. and Phillip, J. (eds.) *Proc. mtg. Les facies de plate-forme du Crétacé dans le domaine Mésogéen, 26th Int. geol. Congr., Paris, 1980, Cret. Res.* **3** (1–2), 3–18.

Arnaud-Vanneau, A. and Medus, J. (1977). Palynoflores barremo-aptiennes de la plate-forme urgonienne du Vercors. Palynostratigraphie de quelques formes de *Classopollis* et de quelques pollens angiospermiens. *Géol. Alpine.* **53**, 35-55.

Arnold, H. (1964). Fazies und Mächtigkeit von Kreidestufen im Münsterland. In: *Die Kreide Westfalens – Ein Symposium. Fortschr. Geol. Rheinl. Westfalen.* **7**, 599–610.

Artru, P. (1972). *Les Terres noires du Bassin Rhodanien (Bajocien supérieur à Oxfordien Moyen): stratigraphie, sédimentologie, géochimie.* Unpubl. Ph.D. Thesis, L'Université Claude Bernard, Lyon, 173pp.

Asmus, H.E. and Campos, D.A. (1983). Stratigraphic division of the Brazilian continental margin and its paleogeographic significance. In: *2nd Symp. Kreide, Munich, 1982. Zitteliana.* **10**, 265–76.

Asmus, H.E. and Ponte, F.C. (1973). The Brazilian marginal basins. In: Nairn, A.E.M. and Stehli, F.G. (eds.) *The ocean basins and margins,* **1**, *The South Atlantic.* Plenum Press, London, 87–134.

Asquith, D.O. (1966). Geology of Late Cretaceous Mesaverde and Paleocene Fort Union oil production, Birch Creek unit, Sublette County, Wyoming. *Bull. Am. Assoc. Petrol. Geol.* **50** (10), 2176–2184.

Asquith, D.O. (1970). Depositional topography and major marine environments, Late Cretaceous, Wyoming. *Bull. Am. Assoc. Petrol. Geol.* **54** (7), 1184–1224.

Asquith, D.O. (1974). Sedimentary models, cycles, and deltas, Upper Cretaceous, Wyoming. *Bull. Am. Assoc. Petrol. Geol.* **58** (11), 2274–2283.

Assereto, R. (1966). The Jurassic Shemshak Formation in Central Elburz (Iran). *Riv. Ital. Palaeontol. Stratigr.* **72** (4), 1133–1182.

Aubague, M., Lefavrais-Raymond, A., L'Homer, A. and Michard, A.G. (1979). Lias des Causses; minéralisations. In: La sédimentation Jurassique W. européen. *Proc. Symp., Paris, May 1977, Assoc. Sédimentol. Fr. Spéc. Publ.* **1**, 235–246.

Aubouin, J., Borrello, A.V., Cecioni, G., Charrier, R., Chotin, P., Frutos, J., Thiele, R. and Vicente, J-C. (1973). Esquisse paléogéographique et structurale des Andes méridionales. *Rev. Géogr. phys. Géol. dyn.* **15** (1–2), 11–72.

Aubouin, J., Sigal, J., Berland, J-P., Blanchet, R., Bonneau, M., Cadet, J-P., Guillot, P-L., Lacour, A., Piat, B. and Vicente, J-C. (1970). Sur un bassin de flysch: stratigraphie et paléogéographie des flyschs Crétacés de la Lombardie (Versant sud des Alpes orientales, Italie). *Bull. Soc. géol. Fr. Sér. 7.* **12** (4), 612–658.

Audebaud, E., Capdevila, R., Dalmayrac, B., Debelmas, J., Laubacher, G., Lefevre, C., Marocco, R., Martinez, C., Mattauer, M., Megard,

F., Paredes, J. and Tomasi, P. (1973). Les traits géologiques essentiels des Andes centrales (Pérou-Bolivie). *Rev. Géogr. phys. Géol. dyn.* **15** (1-2), 73–114.

Audley-Charles, M.G. (1966). Mesozoic palaeogeography of Australia. *Palaeogeogr. Palaeoclimatol. Palaeoecol.* **2** (l), 1–25.

Audley-Charles, M.G. (1970). Triassic palaeogeography of the British Isles. *Q.J. geol. Soc. Lond.* **126** (1), 49–89.

Audley-Charles, M.G. (1977). Mesozoic evolution of the margins of Tethys in Indonesia and the Philippines. *Proc. 5th Ann. Conv. Indonesian Petrol. Assoc., Jakarta 1976.* **2**, 25–52.

Audley-Charles, M.G. (1978). The Indonesian and Philippine Archipelagos. In: Moullade, M. and Nairn, A.E.M. (eds.) *The phanerozoic geology of the world,* **2**, *The Mesozoic, A.* Elsevier, Amsterdam, 165–207.

Audley-Charles, M.G. (1983). Reconstruction of eastern Gondwanaland. *Nature.* **306**, 48–50.

Austin, J.A., Uchupi, E., Shaughnessy, D.R. and Ballard, R.D. (1980). Geology of New England passive margin. *Bull. Am. Assoc. Petrol. Geol.* **64** (4), 501–526.

Austin, P.M. (1975). Paleogeographic and paleotectonic models for the New Zealand geosyncline in eastern Gondwanaland. *Bull. geol. Soc. Am.* **86** (9), 1230–1234.

Azema, J., Bourrouilh, R., Champetier, Y., Fourcade, E. and Rangheard, Y. (1974). Rapports stratigraphiques, paléogéographiques et structuraux entre la Chaîne Ibérique, les Cordillères Bétiques et les Baléares. *Bull. Soc. géol. Fr. Sér. 7.* **16** (2), 140–160.

Azema, J., Champetier, Y., Foucault, A., Fourcade, E. and Paquet, J. (1971). Le Jurassique dans la partie orientale des zones externes des Cordillères Bétiques: essai de coordination. In: Virgili, C. (dir.) *Proc. Coloquio de estratigrafía y paleogeografía del Jurásico d'España, Madrid, 1971, Cuad. Geol. Ibérica.* **2**, 91–110.

Azema, J., Foucault, A. and Fourcade, E. (1979). Le Jurassique des Cordillères Bétiques. In: *La sédimentation Jurassique W. européen. Proc. Symp., Paris, May 1977, Assoc. Sédimentol. Fr. Spéc. Publ.* **1**, 317–334.

Azema, J., Foucault, A., Fourcade, E. and Champetier, Y. (1975). Le Crétacé dans la partie orientale des zones externes des Cordillères Bétiques. In: *Primer coloquio de estratigrafía y paleogeografía del Cretácico de España, Bellaterra-Tremp, November 1973, Publ. Empresa Nac. Adaro. invest. Mineras, Ser. 7, Trabajos de congresos y reuniones., Madrid,* **1**, 159–224.

Azzaroli, A. and Cita, M.B. (1963). *Geologia stratigrafica,* **2**. La Goliardica, Modena, Italy, 353pp. [NB. Not seen; unavailable in U.K. even by BLL Z search; see, however, C.P.75.]

Backhaus, E. (1971). Red beds of the 'Buntsandstein' (Lower Triassic) in the Odenwald and Pfalz-Vosges areas. In: Muller, G. (ed.) *Sedimentology of parts of Central Europe, Guidebook to excursions. 8th Int. Sedimentol. Congr., Heidelburg, 1971.* Verlag Waldemar Kramer, Frankfurt. 105–124.

Backhaus, E. (1974). Limnische und fluviatile Sedimentation in südwestdeutschen Buntsandstein. *Geol. Rdsch.* **63**, 925–942.

Badon, C.L. (1975). Stratigraphy and petrology of Jurassic Norphlet Formation, Clarke County, Mississippi. *Bull. Am. Assoc. Petrol. Geol.* **59** (3), 377–392.

Baeza Assís, L. (1979). Distribución de facies sedimentarias marinas en el Jurásico de Cerritos Bayos y zonas adyacentes-norte de Chile. *Acta. 2nd Congr. Geol. Chileno, 3, Sect. H, Paleontology and stratigraphy,*

Arica-Chile, August 1979, Inst. Investig. Geol. Ciudad de Arica. H45–6l.

Bagshaw, L.H. (1977). Paleoecology of the Lower Carmel Formation of the San Rafael Swell, Emery County, Utah. *Brigham Young Univ. Stud.* **24** (2), 51–62.

Baker, P.G. (1981). Interpretation of the Oolite Marl (Upper Aalenian, Lower Inferior Oolite) of the Cotswolds, England. *Proc. Geol. Assoc. Lond.* **92** (3), 169–187.

Balda, F.A. (1974). Ideas sobre la paleogeografía y paleogeología del oriente y occidente de Venezuela. *Bol. Inst. Oceanogr. Univ. Oriente.* **13** (1–2), 87–106.

Baldwin, E.J. (1973). The Moenkopi Formation of north-central Arizona: an interpretation of ancient environments based upon sedimentary structures and stratification types. *J. sediment. Petrol.* **43** (1), 92–106.

Balkwill, H.R. (1978). Evolution of Sverdrup basin, Arctic Canada. *Bull. Am. Assoc. Petrol. Geol.* **62** (6), 1004–1028.

Balkwill, H.R., Wilson, D.G. and Wall, J.M. (1977). Ringnes Formation (Upper Jurassic), Sverdrup Basin, Canadian Arctic Archipelago. *Bull. Can. Petrol. Geol.* **25**, 1115–1144.

Balkwill, H.R., Cook, D.G., Detterman, R.L., Embry, A.F., Hakansson, E., Miall, A.D., Poulton, T.P. and Young, F.G. (1983). Arctic North America and northern Greenland. In: Moullade, M. and Nairn, A.E.M. (eds.) *The Phanerozoic Geology of the World, 2, The Mesozoic, B.* Elsevier, Amsterdam. 1–31.

Balme, B.E. (1969). The Triassic system in Western Australia. *Aust. Petrol. Explor. Assoc. J.* **9** (2), 67–78.

Balme, B.E. and Helby, R.J. (1973). Floral modifications at the Permian-Triassic boundary in Australia. In: Logan, A. and Hills, L.V. (eds.) *The Permian and Triassic systems and their mutual boundary. Proc. Symp. Univ. Calgary, Aug. 1971. Mem. Can. Soc. Petrol. Geol.* **2**, 433–444.

Bandel, K. (1981). New stratigraphical and structural evidence for lateral dislocations in the Jordan Rift Valley connected with a description of the Jurassic rock column in Jordan. *Neues Jahrb. Geol. Palaontol. Abhandlungen.* **161** (3), 271–308.

Bannerjee, I. (1981). Storm lag and related facies of the bioclastic limestones of the Eze-Aku Formation (Turonian), Nigeria. *Sediment. Geol.* **30** (1), 133–147.

Bannert, D. (1981). The evolution of the Asian plate in Burma. *Geol. Rdsch.* **70** (2), 446–458.

Barber, A.J. (1982). Interpretations of the tectonic evolution of southwest Japan. In: Greensmith, J.T. and Whitaker, J.H.McD. (eds.) *The geology of Japan; Proc. Symp. London, Feb. 1981, Proc. Geol. Assoc. Lond.* **93** (2), 131–145.

Barber, P.M. (1982). Palaeotectonic evolution and hydrocarbon genesis of the Central Exmouth Plateau. *Aust. Petrol. Explor. Assoc. J.* **22** (1), 131–144.

Barfety, J-C. and Mouterde, R. (1980). Evolution des facies du Jurassique dans la zone dauphinoise du Mont Blanc au Pelvoux (Alpes occidentales). *Bull. Soc. géol. Fr. Sér. 7.* **22** (4), 557–565.

Barnard, T., Cordey, W.G. and Shipp, D.J. (1981). Foraminifera from the Oxford Clay (Callovian-Oxfordian of England). *Rev. Esp. Micropaleontol.* **13** (3), 383–462.

Barnes, S.U. (1976). Geology and oil prospects of Somalia, East Africa. *Bull. Am. Assoc. Petrol. Geol.* **60** (3), 389–413.

Barr, F.T. (1972). Cretaceous biostratigraphy and planktonic foraminifera of Libya. *Micropaleontol.* **18** (1), 1–46.

Barrett, P.J., Grindley, G.W. and Webb, P.N. (1972). The Beacon Supergroup of East Antarctica. In: Adie, R.J. (ed.) *Antarctic geology and geophysics, Proc. Symp. Antarctic Geology and Solid Earth Geophysics, Oslo, August 1970, Sci. Comm. Antarctic Research, Int. Union, geol. Sci., Ser. B.* **1**, 319–332.

Barron, E.J., Harrison, C.G.A., Sloan II, J.L. and Hay, W.W. (1981). Paleogeography, 180 million years ago to the present. *Eclog. geol. Helv.* **74** (2), 443–470.

Barss, D.L., Best, E.W. and Meyers, N. (1964). Triassic. In: McGrossan, R.G. and Glaister, R.P. (eds.) *Geological History of Western Canada. Alberta Soc. Petrol. Geol., Calgary.* 113–136.

Barthe, A. and Stevaux, J. (1971). Le bassin à évaporites du Lias inférieur de l'Aquitaine. *Bull. Cent. Rech. Pau.* **5** (2), 363–369.

Bartok, P., Reijers, T.J.A. and Juhasz, I. (1981). Lower Cretaceous Cogollo Group, Maracaibo basin, Venezuela: sedimentology, diagenesis and petrophysics. *Bull. Am. Assoc. Petrol. Geol.* **65** (6), 1110–1134.

Bartov, J., Eyal, Y., Garfunkel, Z. and Steinitz, G. (1972). Late Cretaceous and Tertiary stratigraphy and paleogeography of Southern Israel. *Israel J. earth Sci.* **21** (1), 69–97.

Bartov, Y. and Steinitz, G. (1977). The Judea and Mount Scopus Groups in the Negev and Sinai with trend surface analysis of the thickness data. *Israel J. earth Sci.* **26** (3-4), 119–148.

Basha, S. (1982). Distribution of Triassic rocks in Jordan and Levant. In: *GEOCOME-1, First geological congress of the Middle East, Ankara, Sept. 1979. Min. res. Explor. Inst., Ankara, Turkey.* 179–205.

Bassoullet, J.P. and Mouterde, R. (1977). Les formations sédimentaires Mésozoïques du domaine Tibétain de l'Himalaya du Népal. In: Jest, C. (ed.) *Proc. Colloq. Int. C.N.R.S., Himalaya, Sci. de la Terre.* **268** (2), 53–60.

Bate, R.H. (1967). Stratigraphy and palaeogeography of the Yorkshire oolites and their relationship with the Lincolnshire Limestone. *Bull. Br. Mus. Nat. Hist. Ser. Geol.* **14** (4), 111–141.

Baturin, G.N. (1982). *Phosphorites on the sea floor: origin, composition and distribution. Devs. Sedimentol.* **33**, Elsevier, Amsterdam. 343pp.

Baudrimont, A.F. and Dubois, P. (1977). Un Bassin Mésogéen du domaine péri-Alpin: Le sud-est de la France. *Bull. Cent. Rech. Explor. Prod. Elf-Aquitaine.* **1** (1), 261–308.

Bay, T.A. Jnr. (1977). Lower Cretaceous stratigraphic models from Texas and Mexico. In: Bebout, D.G. and Loucks, R.G. (eds.) *Cretaceous carbonates of Texas and Mexico: applications to subsurface exploration, Rep. Invest. Bur. econ. Geol., Univ. Texas.* **89**, 12–30.

Beauchamp, J. and Lemoigne, Y. (1974). Présence d'un bassin de subsidence en Ethiopie centrale et essai de reconstitution paléogéographique de l'Ethiopie durant le Jurassique. *Bull. Soc. géol. Fr. Sér. 7.* **16** (5), 563–569.

Beauvais, L. (1964). Etude stratigraphique et paléontologique des formations à madréporaires du Jurassique Supérieur du Jura et de l'est du Bassin de Paris. *Mém. Soc. géol. Fr. Hors Sér.* **100**, 288pp.

Beauvais, L. (1974). Quelques exemples pris dans le Malm montrant le rôle de Madréporaires dans la théorie de la Mobilité continentale. *Bull. Soc. géol. Fr. Sér. 7.* **16** (4), 465–469.

Beauvais, L. (1980). Evolution des récifs au cours du Jurassique. *Bull. Soc. géol. Fr. Sér. 7.* **22** (4), 595–598.

Beauvais, L. (1982). Paléobiogéographie des Madréporaires du Trias. In: *Proc. Symp. Biogéographie et tectonique des plaques, Brest, Mar. 1982.*

Bull. Soc. géol. Fr. Sér. 7. **24** (5–6), 963–970.

Bebout, D.G. (1977). Sligo and Hooston depositional patterns, subsurface of South Texas. In: Bebout, D.G. and Loucks, R.G. (eds.) *Cretaceous carbonates of Texas and Mexico: applications to subsurface exploration, Rep. Invest. Bur. econ. Geol. Univ. Texas.* **89**, 79–96.

Bebout, D.G., Schatzinger, R.A. and Loucks, R.G. (1977). Porosity distribution in the Stuart City Trend, Lower Cretaceous, South Texas. In: Bebout, D.G. and Loucks, R.G. (eds.) *Cretaceous carbonates of Texas and Mexico: applications to subsurface exploration, Rep. Invest. Bur. econ. Geol. Univ. Texas.* **89**, 234– 256.

Bechstadt, T. (1978). Faziesanalyse permischer und triadischer Sedimente des Drazugess als Hinweis auf eine grobräumige Lateral verschiebung innerhalb des Ostalpins. *Jahrb. geol. Bundesanst.* **121** (1), 1–21.

Bechstadt, T. and Mostler, H. (1976). Riff-Becken-Entwicklung in der Mitteltrias der westlichen Nördlichen Kalkalpen. *Z. Deutsch. geol. Ges.* **127**, 271–289.

Beck, C. (1983). Essai sur l'évolution géodynamique des Caraïbes sud-orientales. *Bull. Soc. géol. Fr. Sér. 7.* **25** (2), 169–183.

Began, A. and Salaj, J. (1978). Nove paleogeograficke poznatky vo vrchnej kriede a v paleogene zapadneho a stredneho Slovenska. In: Vozar, J. (ed.) *Paleogeographical evolution of the West Carpathians, Proc. Symp. Bratislava, April 1977, Geologicky ustav Dionyza Stura, Bratislava.* 161–174.

Behrens, E.W. (1963). Environment reconstruction for a part of the Glen Rose Limestone, central Texas. *Sedimentol.* **4** (1), 65–111.

Behrens, M. (1972). Schwermineralverteilungen und Sedimentstrukturen in den Lunzer Schichten (Karn, Trias, Österreich). *Jahrb. geol. Bundesanst.* **116** (1), 51–83.

Behrens, M. and Siehl, A. (1982). Sedimentation in the Atlas Gulf I: Lower Cretaceous clastics. In: Rad, U. von, Hinz, K., Sarnthein, M. and Seibold, E. (eds.) *Geology of the Northwest African Continental Margin.* Springer-Verlag. 427–438.

Behrman, P.G. and Parkison, G.A. (1978) Paleogeographic significance of the Callovian to Kimmeridgian strata, central Sierra Nevada foothills, California. In: Howell, D.G. and McDougall, K.A. (eds.) *Mesozoic paleogeography of the western United States, Proc. 2nd Symp. Pacific Coast Paleogeogr, April 1978, Pacific Sect. Soc. econ. Paleontol. Mineral.* 349–360.

Bein, A. (1976). Rudistid fringing reefs of Cretaceous shallow carbonate platform of Israel. *Bull. Am. Assoc. Petrol. Geol.* **60** (2), 258–272.

Bein, A. (1977). Shelf basin sedimentation: mixing and diagenesis of pelagic and clastic Turonian carbonates, Israel. *J. sediment. Petrol.* **47** (1), 382–391.

Bein, A. and Gvirtzman, G. (1977). A Mesozoic fossil edge of the Arabian Plate along the Levant coastline and its bearing on the evolution of the eastern Mediterranean. In: Biju-Duval, B. and Montadert, L. (eds.) *Structural history of the Mediterranean basins. Proc. Int. Symp. Split, Yugoslavia, Oct. 1976 (25th Mtg. C.I.E.S.M.)* Editions Technip, Paris, 95–109.

Belding, H.F. (1955). Geological development of the Colombian Andes. In: Bullard, F.M. (ed.) *Proc. Conf. Latin-American geology, Austin, March 1954, Univ. Texas.* 43–63.

Belford, D.J. (1958). Stratigraphy and micropalaeontology of the Upper Cretaceous of western Australia. *Geol. Rdsch.* **47**, 629–647.

Beloussov, V.V. and Ruditch, E.M. (1961). Island arcs in the development of the earth's structure (especially in the region of Japan and the Sea of Okhotsk). *J. Geol.* **69** (6), 647–658.

Benavides, V.E. (1956). Cretaceous system in Peru. *Bull. Am. Mus. nat. Hist.* **108** (4), 357–393.

Benavides, V. (1968). Saline deposits of South America. In: Mattox, R.B. (ed.) *Saline Deposits. Proc. Symp. Int. Conf. Saline Deposits, Houston, 1962, Spec. Pap. geol. Soc. Am.* **88**, 249–290.

Benbow, D.D. (1980). The petroleum prospects of the Great Barrier Reef Region. *Aust. Petrol. Explor. Assoc. J.* **20** (1), 159–175.

Bender, F. (1968). Geologie von Jordanien. *Beiträge zur regionalen Geologie der Erde.* **7**. Gebrüder Borntraeger, Berlin, 230pp.

Bender, F. (1975). Geology of the Arabian Peninsula, Jordan. *Prof. pap. U.S. geol. Surv.* **560–I**, 136pp.

Bendukidze, N.S. (1977). Ecology of the Malm-Reef formation of the Great Caucasus. In: Second symposium international sur les coraux et récifs coralliens fossiles, *Proc. Symp. Paris, Sept. 1975, Mém. Bur. Rech. géol. minières.* **89**, 313–321.

Benest, M., Dubel, C. and Elmi, S. (1978). Modalités de l'apparition de la sédimentation carbonatée de plate-forme interne sur la frange méridionale du Domaine Tlemcenien pendant l'Aalenien et le Bajocien. 'Les Dolomies du Tenouchfi' (Algérie nord-occidentale). In: *Les sédiments: leur histoire, leur environnement et leur devenir, Livre Jubilaire Jacques Flandrin. Doc. Lab. Géol. Fac. Sci. Lyon. Hors Sér.* **4**, 29–70.

Bennison, G.M. and Wright, A.E. (1969). *The geological history of the British Isles.* Edward Arnold, London. 406pp.

Berberian, M. and King, G.C.P. (1981). Towards a paleogeography and tectonic evolution of Iran. *Can. J. Earth Sci.* **18** (2), 210–265.

Berger, K. (1981). Trias nördlich der Alpen, 3.3, Keuper. In: Haunschild, H. and Jarz, H. (eds.) *Erläuterungen zur Geologischen Karte von Bayern 1 : 500,000.* Bayerisches Geologisches Landesamt, München, 49–54.

Bergner, H.D., Gebhard, G. and Wiedmann, J. (1982). Kondensationserscheinungen in der Marokkanischen und alpinen Mittelkreide (Apt, Alb). In: Wiedmann, J., Kullman, J. and Schonenberg, R. (eds.) *Subsidenz-Entwicklung im Kantabrischen Variszikum und an passiven Kontinentalrändern der Kreide, 2, Kreide, Neues Jahrb. Geol. Paläont. Abhandlungen.* **165** (1), 102–124.

Bergougnan, H. and Fourquin, C. (1980). Un ensemble d'éléments communs à une marge active alpine des Carpathes Méridionales à l'Iran central: le domaine irano-balkanique. *Bull. Soc. géol. Fr. Sér. 7.* **22** (1), 61–83.

Berneis, H.P. (1983). A lower liassic offshore bar environment, contribution to the sedimentology of the Luxembourg sandstone. *Ann. Soc. geol. Belg.* **106** (1), 87–102.

Bernier, P. and Enay, R. (1972). Figures d'émersion temporaire et indices de sédimentation à très fiable profondeur dans le Portlandien et le Kimmeridgien supérieux (Calcaires en plaquettes) du Grand-Colombier-de-Culoz (Ain, France). *Bull. Soc. géol. Fr. Sér. 7.* **14** (1–5), 281–292.

Bernoulli, D., Bischsel, M., Bolli, H.M., Haring, M.O., Hochuli, P.A. and Kleboth, P. (1981). The Missaglia Megabed, a catastrophic deposit in the Upper Cretaceous Bergamo Flysch, northern Italy. *Eclog. geol. Helv.* **74** (2), 421–442.

Bertelsen, F. (1978). The Upper Triassic – Lower Jurassic Vinding and Gassum Formations of the Norwegian-Danish Basin. *Dan. geol. Unders. Ser. B.* **3**, 26pp.

Bertelsen, F. (1980). Lithostratigraphy and depositional history of the Danish Triassic. *Dan. geol. unders. Ser. B.* **4**, 59pp.

Berthou, P.Y. (1978). La transgression Cénomaniennc dans le bassin occidental Portugais. In: Philip, J. (ed.) *Colloque sur le Cénomanien: France-Europe occidentale. Paris, Sept. 1976, Geol. Méditerr., Univ. Provence.* **5** (1), 31–38.

Berthou, P.Y., Blanc, P. and Chamley H. (1982). Sédimentation argileuse comparée au Crétacé moyen et supérieur dans le bassin occidental Portugais et sur la marge voisine (site 398 DSDP): enseignements paléogéographiques et tectoniques. *Bull. Soc. géol. Fr. Sér. 7.* **24** (3), 461–472.

Berthou, P.Y., Ferreira Soares, A., and Lauverjat, J. (1979). Portugal. *Proc. Mid. Cretaceous Events Iberian Field Conf. Madrid, 1977, Guide I, Cuad. Geol. Iberica.* **5**, 31–124.

Berthou, P.Y. and Lauverjat, J. (1978). Essai de reconstitution de l'évolution de la plate-forme Ca carbonatée correspondant au bassin occidental Portugais de l'albien supérieur au Turonien. In: *Evénements de la partie moyenne du Crétacé. Proc. Symp. Sept. 1977, 3, Cah. Micropaléontol.* **1978** (3), 33–49.

Berthou, P.Y. and Lauverjat, J. (1979). Essai de synthèse paléogéographique et paléobiostratigraphique du bassin occidental Portugais au cause du Crétacé supérieur. In: *Sobre o 1st coloquio de estratigrafia e paleogeografia de Meso-Cenozoico Portuguese, Lisboa, April 1978. Cienc. Terra (Fac. Cienc. Tec., Univ. Nova Lisboa).* **5**, 121–144.

Beurlen, K. (1964). Einige Bemerkungen zur erdgeschichtlichen Entwicklung Nordost-Brasiliens. *Neues Jahrb. Geol. Paläontol. Monatshefte.* **1964** (2), 82–105.

Beurlen, K. (1970). Geologie von Brasilien. *Beiträge zur regionalen Geologie der Erde.* **9**, Gebrüder Borntraeger, Berlin, 444pp.

Beurlen, K. (1971). Beiträge zur Palaogeographie der Kreide in Nordost-Brasilien. *Neues Jahrb. Geol. Palaontol. Abhandlungen.* **139** (1), 1–28.

Beurlen, K. (1974). Die geologische Entwicklung des Atlantischen Ozeans. *Geotekton. Forsch.* **46** (1), 1–69.

Beuther, A. (1965). Geologische Untersuchungen in Wealden und Utrillas-Schichten im Westteil der Sierra de los Cameros (Nordwestliche Iberische Ketten). In: Beuther, A., Dahm, H., Kneupe-Haack, F. et al. (1966). *Der Jura und Wealden in Nordost- Spanien. Beih. geol. Jahrb.* **44**, 103–121.

Beutler, G. and Schuler, F. (1979). Über Vorkommen salinarer Bildungen in der Trias im Norden der DDR. *Z. geol. Wiss. DDR.* **7** (7), 903–912.

Beyth, M. (1972). Paleozoic-Mesozoic sedimentary basin of Mekele outlier, northern Ethiopia. *Bull. Am. Assoc. Petrol. Geol.* **56** (12), 2426–2439.

Beznosov, N.V., Gorbatchik, T.N., Mikhailova, I.A. and Pergament, M.A. (1978). Soviet Union. In: Moullade, M. and Nairn, A.E.M. (eds.) *The Phanerozoic geology of the world, 2, The Mesozoic, A.* Elsevier, Amsterdam, 5–54.

Bhat, H., McMillan, N.J., Aubert, J., Porthault, B. and Surin, M. (1975). North American and African drift – the record in Mesozoic coastal plain rocks, Nova Scotia and Morocco. In: Yorath, C.J., Parker, E.R. and Glass, D.J. (eds.) *Canada's continental margins and offshore petroleum exploration. Mem. Can. Soc. Petrol. Geol.* **4**, 375–390.

Bichsel, M. and Haring, M.O. (1981). Facies evolution of Late Cretaceous Flysch in Lombardy (northern Italy). *Eclog. geol. Helv.* **74** (2), 383–420.

Bielecka, W. and Styk, O. (1966). Microfauna malmu południowej części syneklizy perybaityckiej. *Kwart. geol.* **10** (2), 350–366.

Bielecka, W. and Styk, O. (1968). Analiza zespotów mikrofauny oksfordu i kimerydu Polski nizowej w zależności od różnić facjalnych. *Kwart. geol.* **12** (2), 324–344.

Bielecka, W. and Sztejn, J. (1966). Stratygrafia warstw przejściowych między jura a kreda na podstawie mikrofauny. *Kwart. geol.* **10** (1), 96–115.

Bigarella, J.J. (1972). Eolian environments – their characteristics, recognition and importance. In: Rigby, J.K. and Hamblin, W.K. (eds.) Recognition of ancient sedimentary environments. *Spec. Publ. Soc. econ. Paleontol. Mineral.* **16**, 12–62.

Biju-Duval, B., Dercourt, J. and Le Pichon, X. (1977). From the Tethys ocean to the Mediterranean seas: a plate tectonic model of the evolution of the western Alpine system. In: Biju-Duval, B. and Montadert, L. (eds.) *Structural history of the Mediterranean basins. Proc. Int. Symp., Split, Yugoslavia, 1976 (25th Mtg. C.I.E.S.M).* Editions Technip, Paris, 143–164.

Biju-Duval, B. amd Montadert, L. (eds.) (1977). Structural history of the Mediterranean basins. *Proc. Int. Symp., Split, Yugoslavia, 1976, (25th Mtg. C.I.E.S.M.).* Editions Technip, Paris, 448pp.

Bilodeau, W.L. and Lindberg, F.A. (1983). Early Cretaceous tectonics and sedimentation in Southern Arizona, southwestern New Mexico, and northern Sonora, Mexico. In: Reynolds, M.W. and Dolly, E.D. (eds.) *Mesozoic Paleogeography of the West-Central United States. Proc. 2nd Rocky Mt. Paleogeogr. Symp., Denver, Colorado. Rocky Mt. Sect. Soc. econ. Paleontol. Mineral.* 173–188.

Birkenmajer, K. (1964). Lower Cretaceous tidal deposits of central Vestspitsbergen. *Norsk Polarinst. Arbok.* 73–85.

Bischoff, V.G. and Wolburg, J. (1963). Zur Entwicklung der Ober-Malm im Emsland. *Erdöl-Zeit.* **79** (10), 445–472.

Bishop, B.A. (1970). Stratigraphy of Sierra de Pichachos and vicinity, Nuevo Leon, Mexico. *Bull. Am. Assoc. Petrol. Geol.* **54** (7), 1245–1270.

Bishop, W.F. (1967). Age of pre-Smackover formations, north Louisiana and south Arkansas. *Bull. Am. Assoc. Petrol. Geol.* **51** (2), 244–250.

Bishop, W.F. (1968). Petrology of upper Smackover limestone in North Haynesville field, Claiborne Parish, Louisiana. *Bull. Am. Assoc. Petrol. Geol.* **52** (1), 92–128.

Bishop, W.F. (1969). Environmental control of porosity in the Upper Smackover limestone, North Haynesville field, Claiborne Parish, Louisiana. *Trans. Gulf Coast Assoc. geol. Soc.* **19**, 155–169.

Bishop, W.F. (1980). Petroleum geology of northern central America. *J. Petrol. Geol.* **3** (1), 3–59.

Bismuth, H., Bonnefous, J. and Dufaure, P. (1967). Mesozoic microfacies of Tunisia. In: Martin, L. (ed.) *Guide to the geology and history of Tunisia. 9th Ann. Field Conf. Petrol. Explor. Soc. Libya.* Holland-Breumelhof N.V. Amsterdam. 158–214.

Bissell, H.J. (1970). Petrology and petrography of Lower Triassic marine carbonates of southern Nevada (U.S.A.). *Int. sediment. petrogr. Ser.* **14**, E.J. Brill, Leiden, 27pp. with plates.

Bissell, H.J. (1973). Permian-Triassic boundary in the eastern Great Basin area. In: Logan, A. and Hills, L.V. (eds.) *The Permian and Triassic systems and their mutual boundary. Proc. Symp. Univ. 4Calgary, Aug. 1971, Mem. Can. Soc. Petrol. Geol.* **2**, 318–344.

Bitterli, P. (1962). Studien an bituminosen Gesteinen aus Österreich und benachbarten Gebieten. *Erdöl-Zeit.* **78** (7), 405–416.

Bitterli, P.H. (1979). Dogger Jura suisse. In: *La sédimentation Jurassique W. européen. Proc. Symp. Paris, May 1977, Assoc. Sédimentol. Fr. Spéc. Publ.* **1**, 99–110.

Blakey, R.C. and Gubitosa, R. (1983a). Late Triassic paleogeography and depositional history of the Chinle Formation, southern Utah and Northern Arizona. In: Reynolds, M.W. and Dolly, E.D. (eds.) *Mesozoic palaeogeography of West-Central United States, Proc. 2nd Rocky Mt. Paleogeogr. Symp., Denver, Colorado. Rocky Mt. Sect. Soc. econ. Paleontol. Mineral.* 57–76.

Blakey, R.C. and Gubitosa, R. (1984). Controls of sandstone body geometry and architecture in the Chinle Formation (Upper Triassic), Colorado Plateau. In: Nilsen, T.H. (ed.) *Fluvial sedimentation and related tectonic framework, western North America. Sediment. Geol.* **38** (1-4), 51–86.

Blakey, R.C., Peterson, F., Caputo, M.V. and Voorhees, B.J. (1983b). Paleogeography of Middle Jurassic continental, shoreline, and shallow marine sedimentation, southern Utah. In: Reynolds, M.W. and Dolly, E.D. (eds.) *Mesozoic Paleogeography of the West-Central United States. Proc. 2nd Rocky Mt. Paleogeogr. Symp., Denver, Colorado. Rocky Mt. Sect. Soc. econ. Paleontol. Mineral.* 77–100.

Blanc, J-J. and Guieu, G. (1965). L'urgonien du massif de Puget (sud-est de Marseille). *Bull. Soc. géol. Fr. Sér. 7.* **7**, 593–599.

Blant, G. (1973). Structure et paléogéographie du littoral méridional et oriental de l'Afrique. In: Blant, G. (ed.) *Sedimentary Basins of the African Coasts,* **2**, *South and East Coasts. Assoc. Afr. Geol. Surv.* 193–231.

Blaszwiewicz, A. and Cieslinski, S. (1979). Prace nad systematyżowaniem stratygrafii górnej kredy Polski (poza Karpatami i Sudetami). *Kwart. geol.* **23** (3), 639–645.

Blendinger, W. (1983). Anisian sedimentation and tectonics of the M. Pore - M. Cenera area (Dolomites). *Riv. Ital. Paleontol. Stratigr.* **89** (2), 175–208.

Bloos, G. (1976). Untersuchungen über Bau und Entstehung der feinkörnigen Sandsteine des Schwarzen Jura (Hettangium u.tiefstes Sinemurium) im schwäbischen Sedimentationsbereich. *Arb. Inst. Geol. Paläont. Univ. Stuttgart, N.F.* **71**, 1–269.

Boccaletti, M., Gocev, P. and Manetti, P. (1974). Mesozoic isopic zones in the Black Sea region. *Boll. Soc. geol. Ital.* **93** (3), 547–565.

Boccaletti, M. and Manetti, P. (1972). Caratteri sedimentologici del Calcare Massiccio della Toscana a sud dell'Arno. *Boll. Soc. geol. Ital.* **91** (3), 559–582.

Boigk, H. (1961). Ergebnisse und Probleme stratigraphisch-päläogeographischer Untersuchungen im Buntsandstein Nordwestdeutschlands. *Geol. Jahrb.* **78**, 123–134.

Boigk, H., Dietz, C., Grahle, H-O., Hoffmann, K., Hollstein, W., Kuhne, F., Richter, W., Schneeklotn, H. and Wager, R. (1960). Zur Geologie des Emslandes. *Beih. geol. Jahrb.* **37**, 419pp.

Boirie, J-M. and Souquet, P. (1982). Les Poundingues de Mendibelza: dépôts de cônes sous-marins du rift Albien des Pyrénées. *Bull. Cent. Rech. Explor. Prod. Elf-Aquitaine.* **6** (2), 405–435.

Bonaparte, J.F. and Powell, J.E. (1980). A continental assemblage of tetrapods from the Upper Cretaceous beds of El Brete, northwestern Argentina (Savropoda-Ceolurosauria-Carnasauria-Ares). In: *Ecosystèmes continentaux du Mésozoique. Proc. Symp. C.N.R.S., Paris, Sept. 1978. Mém. Soc. géol. Fr. Hors Sér.* **59** (193), 19-28.

Bonneau, M. (1969). Contribution à l'étude géologique de la Judicarie, au Nord-Ouest du lac de Garde (Alpes méridionales, province de Trente, Italie). *Bull. Soc. géol. Fr. Sér. 7.* **11** (6), 816–829.

Bonnefous, J. (1967). Jurassic stratigraphy of Tunisia: a tentative synthesis. In: Martin, L. (ed.) *Guide to the geology and history of Tunisia. 9th Ann. Field Conf. Petrol. Explor. Soc. Libya.* Holland-Breumelhof N.V. Amsterdam. 109–130.

Bortolotti, V., Passerini, P., Sagri, M. and Sestini, G. (1970). The miogeosynclinal sequences. In: Sestini, G. (ed.) *Development of the northern Apennines geosyncline. Sediment. geol.* **4** (3–4), 341–444.

Bosellini, A., Broglio Loriga, C. and Busetto, C. (1978). I bacini Cretacei del Trentino. *Riv. Ital. Paleontol. Stratigr.* **84** (4), 897–946.

Bosellini, A., Masetti, D. and Sarti, M. (1981). A Jurassic 'Tongue of the ocean' infilled with oolitic sands: the Belluno Trough, Venetian Alps, Italy. In: Cita, M.B. and Ryan, W.B.F. (eds.) *Carbonate platforms of the passive-type continental margins: present and past. Mar. Geol.* **44** (1–2), 59–96.

Bottke, H., Dengler, H., Finkenwirth, A., Gruss, H., Hoffman, K., Kolbe, H., Simon, P. and Theinhaus, R. 1969. Sammelwerk Deutsche Eisenerzlagerstätten: II Eisenerze im Deckgebirge (Postvarisstikum) I. Die Marin-sedimentaren Eisenerze des Jura in Nordwestdeutschland. *Beih. geol. Jahrb.* **79**, 391pp.

Boudouresque, L., Dubois, D., Lang, J. and Trichet, J. (1982). Contribution à la stratigraphie et à la paléogéographie de la bordure occidentale du bassin des Iullemmeden au Crétacé supérieur et au Cénozoïque (Niger et Mali, Afrique de l'Ouest). *Bull. Soc. géol. Fr. Sér. 7.* **24** (4), 685–695.

Boullier, A. (1976). Les Térébratulides de l'Oxfordien du Jura et de la Bordure sud du Bassin de Paris. *Ann. sci. Univ. Besancon., 3rd Sér., Géol.* **27**, 457pp.

Bourbon, M., Caron, J-M. and de Graciansky, P-Ch. (1977). Mesozoic evolution of the western Alps: birth and development of the spreading oceanic Tethys and its European continental margin. In: Biju-Duval, B. and Montadert, L. (eds.) *Structural history of the mediterranean basins, Proc. Int. Symp., Split, Yugoslavia, 1976, (25th Mtg. C.I.E.S.M.)* Editions Technip, Paris, 19–34.

Bourbon, M., de Graciansky, P-C. and Roux, M. (1980). Indices bathymétriques fournis par les crinoides pédoncules sur le bord de la marge téthysienne (Brianconnais et sub-brianconnais au Jurassique et au Crétacé). *Bull. Soc. géol. Fr. Sér. 7.* **22** (5), 713–718.

Bouroullec, J., Delfaud, J. and Deloffre, R. (1979). Organisation sédimentaire et paléoécologique de l'Aptien supérieur à facies urgonien dans les Pyrénées occidentales et l'Aquitaine méridionale (France). In: *L'Urgonien des Pays Méditerranéens. Géobios. Spéc. Mém.* **3**, 25–44.

Bouroullec, J., Delfaud, J., Gauthier, J. and Lenguin, M. (1973). Etude sédimentologique de la plate-forme interne carbonatée du Quercy (S.W. France), du Bathonien au Callovien. *Bull. Cent. Rech. Pau.* **7** (2), 437–497.

Bouroullec, J. and Deloffre, R. (1970). Interprétation sédimentologique et paléogéographique, par microfacies, du Crétacé Inférieur basal d'Aquitaine sud-ouest. *Bull. Cent. Rech. Pau.* **4** (2), 381–429.

Bouroullec, J. and De Loffre, R. (1976). Relations facies – environnement au Crétacé Moyen en Aquitaine occidentale. *Bull. Cent. Rech. Pau.* **10** (2), 535–583.

Bourseau, J-P and Elmi, S. (1980). Le passage des facies de bordure (Calcaires grumeleux) aux facies de bassin dans l'Oxfordien de la

bordure vivaro-cévenole du Massif Central francais (Ardèche-Gard). *Bull. Soc. géol. Fr. Sér. 7.* **22** (4), 607–611.

Bowen, R. (1961). Paleotemperature analyses of Mesozoic Belemnoidea from Australia and New Guinea. *Bull. geol. Soc. Am.* **72** (5), 769–774.

Boyd, D.W. and Maughan, E.K. (1973). Permian-Triassic boundary in the Middle Rocky Mountains. In: Logan, A. and Hills, L.V. (eds.) *The Permian and Triassic systems and their mutual boundary. Proc. Symp. Univ. Calgary, Aug. 1971, Mem. Can. Soc. Petrol. Geol.* **2**, 294–317.

Bradshaw, M.J. (1975). Origin of montmorillonite bands in the Middle Jurassic of Eastern England. *Earth Planet. Sci. Lett.* **26**, 245–252.

Bradshaw, M.J. and Bate, R.H. (1982). Lincolnshire borehole proves greater extent of the Scarborough Formation (Jurassic:Bajocian). *J. Micropalaeontol.* **1** (1), 141–147.

Bragin, Y.N., Golubev, S.A. and Polyanskii, B.V. (1981). Paleogeography of major accumulation stages of lower Mesozoic coal deposits in Iran. *Lithol. Min. Resour.* **16** (1), 50–59.

Brand, E. and Hoffman, K. (1963). Stratigraphy and facies of the northwest German Jurassic and genesis of its oil deposits. *Proc. 6th World Petrol Congr., Frankfurt am Main, Sect. 1, Pap.* **17**, 233–246.

Brandner, R. (1984). Meeresspiegelschwankungen und Tektonik in der Trias der N.W. Tethys. *Jahrb. geol. Bundesanst.* **126** (4), 435–476.

Brennand, T.P. (1975). The Triassic of the North Sea In: Woodland, A.W. (ed.) *Petroleum and the Continental Shelf of North West Europe*, **1**, *Geology.* Applied Science, London, 295–310.

Brenner, K. and Seilacher, A. (1978). New aspects about the origin of the Toarcian Posidonia Shales. *Neues Jahrb. Geol. Palaontol. Abhandlungen.* **157** (1), 11–18.

Brenner, R.L. (1983). Late Jurassic tectonic setting and paleogeography of Western Interior, North America. In: Reynolds, M.W. and Dolly, E.D. (eds.) *Mesozoic Paleogeography of the West- Central United States. Proc. 2nd Rocky Mt. Paleogeogr. Symp., Denver, Colorado. Rocky Mt. Sect. Soc. econ. Paleontol. Mineral.* 119–132.

Brenner, R.L. and Davies, D.K. (1974). Oxfordian sedimentation in Western Interior United States. *Bull. Am. Assoc. Petrol. Geol.* **58** (3), 407–428.

Brenner, R.L. (1978). Sussex Sandstone of Wyoming – example of Cretaceous offshore sedimentation. *Bull. Am. Assoc. Petrol. Geol.* **62** (2), 181–200.

Brenner, R.L. (1980). Construction of process-response models for ancient epicontinental seaway depositional systems using partial analogs. *Bull. Am. Assoc. Petrol. Geol.* **64** (8), 1223–1244.

Bridges, P.H. (1982). Ancient offshore tidal deposits. In: Stride, A.H. (ed.) *Offshore tidal sands: processes and deposits.* Chapman and Hall. 172–192.

Brinkman, R. (1971). The geology of Western Anatolia. In: Campbell, A.S. (ed.) *Geology and history of Turkey, Petrol. Explor. Soc. Libya, Tripoli.* 171–190.

Brinkman, R., Gumus, H., Plumhoff, F. and Salah, A.A. (1977). Höhere Oberkreide in Nordwest-Anatolien und Thrakien (Beitrag zur Geologie von Anatolien Nr. 18. *Neues Jahrb. Geol. Paläontol. Abhandlungen.* **154** (1), 1–20.

Brognon, G.P. and Verrier, G.R. (1966). Oil and geology in Cuanza basin of Angola. *Bull. Am. Assoc. Petrol. Geol.* **50 (1), 108–158.**

Brookfield, M.E. (1973). Palaeogeography of the Upper Oxfordian and Lower Kimmeridgian (Jurassic) in Britain. *Palaeogeogr. Palaeoclimatol. Palaeoecol.* **14** (2), 137–167.

Broquet, P. and Thomas, M. (1979). Quelques charactères géologiques et géochimiques des schistes bitumineux du Toarcien franc-comtois. *Bull. Cent. Rech. Explor. Prod. Elf-Aquitaine.* **3** (1), 265–280.

Brotzen, F. (1966). Faziesveränderungen in der Oberkreide Schwedens. In: Wehrl, H. (ed.) *Proc. Int. Symp. Die Oberkreide der Baltischen Strasse. Ber. Deutsch. Ges. geol. Wiss., Ser. A, Geol. Paläeontol.* **11** (6), 709–719.

Brown, C.M., Pigram. C.J. and Skwarko, S.K. (1980). Mesozoic stratigraphy and geological history of Papua New Guinea. *Palaeogeogr. Palaeoclimatol. Palaeoecol.* **29** (3–4), 301–322.

Brown, D.A., Campbell, K.S.W. and Crook, K.A.W. (1968). *The geological evolution of Australia and New Zealand.* Pergamon press, Sydney, 409pp.

Bruhn, R.L. and Dalziel, I.W.D. (1977). Destruction of the Early Cretaceous marginal basin in the Andes of Tierra Del Fuego. In: Talwani, M. and Pitman III, W.C. (eds.) *Island Arcs, Deep Sea Trenches and Back-Arc Basins, Am. geophys. Union, Maurice Ewing Ser.* **1**, 395–406.

Brun, L. and Canerot, J. (1979). *Torremiroella hispanica* n. gen. n. sp. – Un nouveau Lituolide (Foraminifère) du Barremien supérieur du Maestrazgo (Espagne). *Bull. Cent. Rech. Explor. Prod. Elf-Aquitaine.* **3** (2), 311–335.

Bryant, W.R., Meyerhoff, A.A., Brown, N.K., Furrer, M.A., Pyle, T.E. and Antoine, J.W. (1969). Escarpments, reef trends and diapiric structures, eastern Gulf of Mexico. *Bull. Am. Assoc. Petrol. Geol.* **53** (12), 2506–2542.

Buday, T. and Tyracek, J. (1980). The regional geology of Iraq, 1, stratigraphy and paleogeography. *State Organisation for Minerals, Baghdad.* Dar Al-Kutib, Univ. Mosul, Iraq. 445pp.

Bulle, J. (1973). Le Trias dans la région de Titov Veles (zone du Vardar – Macédoine Yougoslavie). *Ann. sci. Univ. Besancon, Sér. 3, Géol.* **18**, 97–104.

Bulle, J., Martin, J. and Rollet, M. (1968). Données actuelles sur les indices d'influence continentale dans le Jurassique Supérieur du Jura. In: Rat. P. (ed.) *Sud-est du bassin de Paris et ses abords à l'ère Sécondaire Bull. Soc. géol. Fr. Sér. 7.* **10** (3), 370–375.

Burger, D. (1966). Palynology of Uppermost Jurassic and Lowermost Cretaceous strata in the eastern Netherlands. *Leidse geol. Meded.* **35**, 209–276.

Burger, D. (1982). A basal Cretaceous dinoflagellate suite from northeastern Australia. *Palynology.* **6**, 161–192.

Burger, H. and Strasser, A. (1981). Lithostratigraphische Einheiten der untersten Helvetischen Kreide in der Zentral-und Ostschweiz. *Eclog. geol. Helv.* **74** (2), 529–560.

Burgess, C.J. and Lee, C.W. (1978). The development of a Lower Jurassic carbonate tidal flat, central High Atlas, Morocco. 1: sedimentary history. *J. sediment. Petrol.* **48** (3), 777–794.

Burgess, I.C. and Holliday, D.W. (1974). The Permo-Triassic rocks of the Hilton borehole, Westmorland. *Bull. geol. Surv. G.B.* **46**, 1–34.

Burgl, H. (1973). *Precambrian to Middle Cretaceous stratigraphy of Colombia.* Translation: Allen, C.G. and Rowlinson, N.R. Bogota, 214pp.

Burnhill, T.J. and Ramsay, W.V. (1980). Mid-Cretaceous palaeontology and stratigraphy, central North Sea. In: Illing, L.V. and Hobson, G.D. (eds.) *Petroleum geology of the continental shelf of N.W. Europe.*

Proc. 2nd conference, London, March 1980, Inst. Petrol. Heydon and Sons. 245–254.

Burollet, P.F. and Desforges, G. (1982). Dynamique des bassins Néocrétacés en Tunisie. In: Lang, J. (ed.) *Livre Jubilaire Gabriel Lucas: géologie sédimentaire. Mém. Géol. Univ. Dijon.* **7**, 381–9.

Burollet, P.F. and Manderscheid, G. (1965). Le Crétacé Inférieur en Tunisie et en Libye. *Colloque sur le Crétacé Inférieur, Lyon, Sept. 1963, Mém. Bur. Rech. géol. minières.* **34**, 785–794.

Burollet, P.F., Memmo, L. and M'Rabet, A. (1983). Le Crétacé Inférieur de Tunisie – apercu stratigraphique et sédimentologique. In: *2nd Symp. Kreide, Munich, 1982. Zitteliana.* **10**, 255–264.

Burollet, P.F. and Winnock, E. (1977). Cartes sédimentologiques et paléogéographiques des régions péri-Méditerranéennes occidentales. In: Biju-Duval, B. and Montadert, L. (eds.) *Structural history of the Mediterranean basins. Proc. Int. Symp., Split, Yugoslavia, Oct. 1976 (25th Mtg. C.I.E.S.M.)* Editions Technip, Paris. 111–27.

Buser, H. (1966). Paleostructures of Nigeria and adjacent countries. *Geotekton. Forsch.* **24**, 1–90.

Bushaw, D.J. (1968). Environmental synthesis of the East Texas Lower Cretaceous. *Trans. Gulf Coast Assoc. geol. Soc.* **18**, 416- 438.

Busnardo, R. (1979). Prébétique et subbétique de Jaen à Lucena (Andalousie) le Lias. *Doc. Lab. Géol. Fac. Sci.* **74**, 140pp.

Busson, G. (1969). Sédimentation, transgression et paléogéographie sur les grandes plates-formes du Mésozoïque: l'exemple du Cénomanien – Turonien du Nord-Est de la plate-forme saharienne et de Berberie. *Bull. Soc. géol. Fr. Sér. 7.* **11** (5), 687–703.

Busson, G. (1970). Le Mésozoïque Saharien, 2, Essai de synthèse des données des sondages algéro-tunisiens. *Cent. Rech. Zones arides (C.N.R.S.), Paris, ser. Geol.* **11**, 811pp.

Busson, G. (1982). Le Trias Comme période salifère. In: Proc. Symp. Triassic global geology, Würzburg, Feb. 1982. *Geol. Rdsch.* **71** (3), 857–880.

Busson, G. (1982). Les observations faites sur les marais salants actuels sont-elles transposables aux grandes accumulations évaporitiques du passé? In: Lang, J. (ed.) *Livre Jubilaire Gabriel Lucas: géologie sédimentaire. Mém. Géol. Univ. Dijon.* **7**, 21–35.

Butterlin, J. (1983). The Caribbean region. In: Moullade, M. and Nairn, A.E.M. (eds.) *The Phanerozoic geology of the world, 2, The Mesozoic, B.* Elsevier, Amsterdam, 89–119.

Cabrit, J.P. (1971). Les dépôts d'évaporites dans le Crétacé Inférieur du bassin d'Aquitaine. *Bull. Cent. Rech. Pau.* **5** (2), 383–388.

Cadet, J-P. (1978). Essai sur l'évolution Alpine d'une paléomarge continentale: Les confins de la Bosnie-Herzegovine et du Montenegro, Yougoslavie. *Mém. Soc. géol. Fr. Hors Sér.* **133**, 83pp.

Callomon, J., Donovan, D.T. and Trumpy, R. (1972). An annotated map of Permian and Mesozoic formations of East Greenland. *Medd. Gronland. udgiune af kommissionen for Vidensk. Undersog. 1. Greenland.* **168**, 33pp.

Callomon, J.H. (1979). Marine boreal Bathonian fossils from the northern North Sea and their palaeogeographical significance. *Proc. Geol. Assoc. Lond.* **90** (4), 163–169.

Camacho, H.H. (1967). Las transgresiones del Cretácico Superior y Terciario de la Argentina. *Rev. Assoc. geol. Argentina.* **22** (4), 253–280.

Campbell, C.J. and Burgl, H. (1965). Section through the eastern cordillera of Colombia, South America. *Bull. geol. Soc. Am.* **76** (5), 567–590.

Campbell, F.A. and Lerbekmo, J.F. (1963). Mineralogic and chemical variations between Upper Cretaceous continental Belly River Shales and marine Wapiabi Shales in western Alberta, Canada. *Sedimentol.* **2**, 215–226.

Campos, C.W.M., Miura, K. and Reis, L.A.N. (1975). The east Brazilian continental margin and petroleum prospects. In: *Proc. 9th World Petrol. Congr.* **2**, *Geology.* Applied Science, London. 71–81.

Campos, C.W.M., Ponte, F.C. and Miura, K. (1974). Geology of the Brazilian continental margin. In: Burk, C.A. and Drake, C.L. (eds.) *The geology of continental margins.* Springer-Verlag, Berlin, 447–462.

Canerot, J. (1979a). Les Algues et leur environnement dans le Malm et le Crétacé Inférieur des Chaînes Ibérique et Catalane (Espagne). *Bull. Cent. Rech. Explor. Prod. Elf-Aquitaine.* **3** (2), 505–517.

Canerot, J. (1979b). Le complexe urgonien dans les chaines Ibérique et Catalane (Espagne). In: *L'Urgonien des Pays Méditerranéens. Géobios. Spéc. Mém.* **3**, 45–56.

Canerot, J. and Cugny, P. (1982). Le plate-forme Hauterivienne des Ibérides sud-orientales (Espagne) et ses environnements biosédimentaires. In: Cotillon, P. and Phillip, J. (eds.) *Proc. Mtg. Les facies de plate-forme du Crétacé dans le domaine Mésogéen, 26th Int. geol. Congr., Paris, 1980, Cret. Res.* **3** (1–2), 91–102.

Canuti, P., Fazzuoli, M., Ficcarelli, G. and Venturi, F. (1983). Occurrence of Liassic faunas at Waaney (Uanei) Province of Bay, south-western Somalia. *Riv. Ital. Palaeontol. Stratigr.* **89** (1), 31–46.

Carbone, F., Praturlon, A. and Sirna, G. (1971). The Cenomanian shelf - edge facies of Rocca di Cave. *Geologica Romana.* **10**, 131–198.

Carey, S.McD. and Colburn, I.P. (1978). Late Cretaceous sedimentation in the Santa Monica Mountains, California. In: Howell, D.G. and McDougall, K.A. (eds.) *Mesozoic paleogeography of the western United States, proc. 2nd Symp. Pacific Coast Paleogeogr, April 1978, Pacific Sect. Soc. econ. Paleontol. Mineral.*, 547–558.

Carlson, C.E. (1968). Triassic-Jurassic of Alberta, Saskatchewan, Manitoba, Montana and North Dakota. *Bull. Am. Assoc. Petrol. Geol.* **52** (10), 1969–1983.

Carozzi, A.V. (1981). Porosity models and oil exploration of Amapa carbonates, Paleogene, Foz du Amazonas Basin, offshore NW. Brazil. *J. Petrol. Geol.* **4** (1), 3–34.

Carr, T.R. and Paull, R.K. (1983). Early Triassic stratigraphy and paleogeography of the Cordilleran Miogeocline. In: Reynolds, M.W. and Dolly, E.D. (eds.) *Mesozoic palaeogeography of the West-Central United States, Proc. 2nd Rocky Mt. Paleogeogr. Symp., Denver, Colorado. Rocky Mt. Sect. Soc. econ. Paleontol. Mineral.* 39–56.

Carrasco, V.B. (1971). Litofacies de la Formación El Abra en la Plataforma de Actopan, Hgo. *Rev. Inst. Mex. Pet.* **3** (1), 5–26.

Carrasco V.B. (1977). Albian sedimentation of submarine autochthonous and allochthonous carbonates, east edge of the Valles-San Luis Potosi Platform, Mexico. In: Cook, H.E. and Enos, P. (eds.) *Deep water carbonate environments. Spec. Publ. Soc. econ. Paleontol. Mineral.* **25**, 263–272.

Carrigy, M.A. (1967). Some sedimentary features of the Athabasca oil sands. *Sediment. Geol.* **1** (4), 327–352.

Carrigy, M.A. (1971). Deltaic sedimentation in Athabasca tar sands. *Bull. Am. Assoc. Petrol. Geol.* **55** (8), 1155–1169.

Casamiquela, R.M. (1978). La zona litoral de la transgresión Maastrichtense en el norte de la Patagonia: Aspectos ecológicos. *Ameghiniana.* **15** (1–2), 137–148.

Cassinis, G. (1968). Stratigrafia e tettonica dei terreni Mesozoici compresi tra Brescia e Serle. *Atti. Ist. geol. Univ. Pavia.* **19**, 50–152.

Cassinis, G. (1979). Punto sulle conoscenze del Giurassico bresciano e relative considerazioni. *Atti Ist. geol. Univ. Pavia.* **27–28**, 36–68.

Cassoudebat, M. and Platel, J-P. (1976). Sédimentologie et paléogéographie du Turonien de la bordure septentrionale du Bassin du aquitain. *Bull. Bur. Rech. géol. minières. Sér. 2, Sect. I,* **2**, 85–102.

Castellarin, A. (1980). Italie: Géologie historique et paléogéographique, Jurassique et Crétacé. In: *Géologie des pays européens: Espagne, Grèce, Italie, Portugal, Yougoslavie. Com. Nat. Fr. géol., 26th Int. geol. congr., Paris 1980.* 237–242.

Cathcart, J.B. and Zambrano, F. (1967). Roca fosfática en Colombia. *Bol. Geol. Minist. Minas Colombia.* 15 (1-3), 65–162.

Cavallin, A. and Martinis, B. (1977). Inquadramento geologico. *Riv. Ital. Paleontol. Stratigr.* **83** (2), 219–235.

Cavallin, A. and Pirini Radrizzani, C. (1983). Evoluzione del settore settentrionale della piattaforma carbonatica Adriatica. *Riv. Ital. Paleontol. Stratigr.* **89** (2), 223–242.

Cavelier, C., Megnien, C., Pomerol, C. and Rat, P. (1980). Bassin de Paris. In: Lorenz, C (ed.) *Géologie des Pays européens: France, Belgique, Luxembourg. 26th Int. geol. Congr., Paris.* p.431–484.

Cecioni, G.O. (1957). Cretaceous flysch and molasse in Departamento Ultima Esperanza, Magallanes Province, Chile. *Bull. Am. Assoc. Petrol. Geol.* **41** (3), 538–564.

Cecioni, G. (1970). Esquema de paleogeografía Chilena. *Colección Recursos Naturales.* **1**, Editorial Universitaria, s.a., Santiago, 144pp.

Cediel, F. (1968). El Grupo Girón: una molasa Mesozoica de la Cordillera Oriental. *Bol. Geol. Minist. Minas. Colombia.* **16** (1–3), 5–96.

Chamney, T.P. (1973). Micropalaeontological correlation of the Canadian boreal Lower Cretaceous. In: Casey, R. and Rawson, P.F. (eds.) *The Boreal Lower Cretaceous, Geol. J. Spec. Issue.* **5**, Seel House Press, Liverpool, 19–40.

Champetier, Y. and Moullade, M. (1970). Corrélations des facies à Orbitolinidae à l'Est du Rio Albaida, provinces de Valence et d'Alicante (Espagne). Conséquences stratigraphiques, paléoécologiques et paléogéographiques. *Bull. Soc. géol. Fr. Sér. 7.* **12** (5), 765–773.

Channell, J.E.T., D'Argenio, B. and Horrath, F. (1979). Adria, the African promontory in Mesozoic Mediterranean palaeogeography. *Earth Sci. Rev.* **15** (3), 213–292.

Chapa, A.C. (1963). Etude biostratigraphique des ammonites du centre et de l'est du Mexique (Jurassique Supérieur et Crétacé). *Mém. Soc. géol. Fr. Hors Sér.* **99**, 103pp.

Chapa, A.C. (1976). Nuevas localidades del Kimeridgiano y Titoniano en Chihuahua (Norte de México). *Rev. Inst. Mex. Pet.* **8** (1), 38–49.

Charrier, R. (1979). El Triasico en Chile y regiones adyacentes de Argentina: una reconstrucción paleogeográfica y paleoclimática. *Commun. Dept. Geol. Univ. Chile.* **26**, 1–37.

Chauve, P., Fluck, P., Enay, R., Sittler, C. and Edel, J.P. (1980). Vosges-Fosse Rhenan Bresse-Jura. In: Lorenz, C. (ed.) *Géologie des Pays européens: France, Belgique, Luxembourg. 26th Int. geol. Congr.* 353–430.

Chemberski, H. and Vaptsarova, A. (1975). Certain features of the lithofacies and palaeogeography of a part of the Middle Triassic in northern Bulgaria. *Palaeontol. Stratigr. Lithol.* **3** (1), 47–59. (In Russian, English summary and bilingual captions).

Chen Pei-ji (1983). A survey of the non-marine Cretaceous in China. *Cret. Res.* **4** (2), 123–143.

Chenet, P.Y. (1979). Le secteur subbrianconnais entre Gap et le col du Galibier au Mésozoïque, sa place dans la marge européenne de la Téthys. *Géol. Alpine.* **55**, 45–74.

Cherven, V.B. (1978). Fluvial and deltaic facies in the Sentinel Butte Formation, central Williston Basin. *J. sediment. Petrol.* **48** (1), 159–170.

Chong, Diaz, G. (1976). El sistema Jurásico en la Cordillera de Domeyko (Chile) entre 24° 30′ y 25° 30′ de lat. sur. In: *Memoria segundo congr. Latinoamericano Geol. November 1973, 2, Publ. esp. Bol. Geol. Caracas.* **7**, 765–785.

Chotin, P. (1976). Essai d'interprétation du Bassin Andin chilieno-argentin mésozoïque en tant que bassin marginal. *Ann. Soc. géol. Nord.* **96** (3), 177–184.

Chotin, P. (1977). Les Andes Méridionales à la latitude de Concepcion (Chile, 38° S): portion intracratonique d'une chaîne developpée en bordure de la marge active est-Pacifique. *Rev. Géogr. phys. Géol. dyn.* **19** (4), 353–376.

Choubert, G. and Favre-Muret, A. (1962). Evolution du domaine Atlasique Marocain depuis les temps Paléozoïques. In: Durand-Delga, M. (ed.) *Livre à la mémoire du Professeur Paul Fallot: L'évolution paléogéographique et structurale des domaines Méditerranéens et Alpins d'Europe. Mém. Soc. géol. Fr. Hors Sér.1* **1**, 447–527.

Christensen, W.K. (1982). Late Turonian – early Coniacian belemnites from western and central Europe. *Bull. geol. Soc. Denmark.* **31** (1–2), 63–79.

Cieslinski, S. (1965). Beitrag zur Kenntnis des Oberkreide-Klimas in Polen. In: *Proc. Symp. 'Palaoklima', Köln, March 1964. Geol. Rdsch.* **54** (1), 318–328.

Cieslinski, S. (1976). Rozwój bruzdy dunsko-polskiej na obszarze Świętokrzyskim w albie, cenomanie i turnie dolnym. *Biul. Inst. geol.* **295**, 249–271.

Ciric, B. (1965). Sur les flyschs et les molasses du cycle alpin dans les Dinarides Yougoslaves. *Bull. Soc. géol. Fr. Sér. 7.* **7**, 499–510.

Cita, M.B. (1965). Jurassic, Cretaceous and Tertiary microfacies from the southern Alps (northern Italy). *Int. sediment. petrogr. Ser.* **8**. E.J. Brill, Leiden, 99pp + plates.

Cita, M.B., Benelli, F., Bigioggero, B., Chezar, H., Colombo, A., Sestini, N.F., Freeman-Lynde, R., Iaccarino, S., Jadoul, F., Legnani, E., Malinverno, A., Massiotta, P., Paggi, L. and Silva, I.P. (1980). Contribution to the geological exploration of the Malta escarpment (eastern Mediterranean). *Riv. Ital. Paleontol. Stratigr.* **86** (2), 317–356.

Cita, M.B. and Premoli Silva, I. Il limite Cretaceo-Paleocene e l'evoluzione della Tetide. In: *Contributi stratigrafici e paleogeografici sul Mesozoico della Tetide. Mem. Riv. Ital. Paleontol. Stratigr.* **14**, 193–249.

Clark, D.L. (1957). Marine Triassic stratigraphy in eastern Great Basin. *Bull. Am. Assoc. Petrol. Geol.* **41** (10), 2192–2222.

Claus, C-D., Grebe, H., Lenteritz, K. and Wirth, W. (1978). Zur Altersstellung und paläogeographischen Bedeutung des Paläokarstes auf der Warsteiner Carbonatplattform. *Neues Jahrb. Geol. Paläontol. Monatshefte.* **1978** (10), 577–589.

Clemens, W.A. (1980). Rhaeto-Liassic Mammals from Switzerland and West Germany. *Zitteliana.* **5**, 52-92.

Clemmensen, L.B. (1978). Lacustrine facies and stromatolites from the Middle Triassic of East Greenland. *J. sediment. Petrol.* **48** (4), 1111–1128.

Clemmensen, L.B. (1980). Triassic rift sedimentation and palaeo-

geography of central East Greenland. *Bull. Gronlands. geol. Undersog.* **136**, 72pp.

Clemmensen, L.B. and Surlyk, F. (1976). Upper Jurassic coal-bearing shoreline deposits, Hochstetter Forland, East Greenland. *Sediment. Geol.* **15** (3), 193–211.

Cobban, W.A. and Reeside, J.B. Jnr. (1952). Frontier Formation, Wyoming and adjacent areas. *Bull. Am. Assoc. Petrol. Geol.* **36** (10), 1913–1961.

Cobbing, E.J., Pitcher, W.S., Wilson, J.J., Baldock, J.W., Taylor, W.P., McCourt, W. and Snelling, N.J. (1981). The geology of the Western Cordillera of northern Peru. *Overseas Mem. Inst. geol. Sci. Lond.* **5**, 143pp.

Colacicchi, R., Passeri, L. and Pialli, G. (1975). Evidence of tidal environmental deposition in the Calcare Massiccio Formation (Central Apennines – Lower Lias). In: Ginsburg, R.N. (ed.) *Tidal deposits; a casebook of recent examples and fossil counterparts.* Springer-Verlag, New York, 345–353.

Colin, J.P., Lehmann, R.A. and Morgan, B.E. (1981). Cretaceous and late Jurassic biostratigraphy of the north Celtic Sea Basin, offshore southern Ireland. In: Neale, J.W. and Brasier, M.D. (eds.) *Microfossils from recent and fossil shelf seas, Proc. Mtg. British Micropalaeont. Soc. July 1980, British Micropalaeont. Soc. Ser.* Ellis Horwood, Chichester, 122–155.

Collins, S.E. (1980). Jurassic Cotton valley and Smackover reservoir trends, east Texas, North Louisiana, and south Arkansas. *Bull. Am. Assoc. Petrol. Geol.* **64** (7), 1004–1013.

Collinson, J.W. and Hasenmueller, W.A. (1978). Early Triassic paleogeography and biostratigraphy of the Cordilleran miogeosyncline. In: Howell, D.G. and McDougall, K.A. (eds.) *Mesozoic paleogeography of the western United States, Proc. 2nd Symp. Pacific Coast Paleogeography, April 1978, Pacific Sect. Soc. econ. Paleontol. Mineral.* 175–188.

Colom, G. (1967). Sur l'interprétation des sédiments profonds de la zone géosynclinale Baléare et Subbétique (Espagne). *Palaeogeogr. Palaeoclimatol. Palaeoecol.* **3** (3), 299–310.

Colom, G. and Escandell, B. (1962). L'évolution du géosynclinal Baléare. In: Durand-Delga, M. (ed.) *Livre à la Mémoire du Professeur Paul Fallot: L'évolution paléogéographique et structurale des domaines méditerranéens et Alpins d'Europe. Mém. Soc. géol. Fr., Hors Sér. 1*, 1, 125–136.

Conard-Noireau, M. (1982). Le Turonien de Haute-Provence: répartition des facies et associations de foraminifères planctoniques. In: *Proc. Colloque sur le Turonien, Paris, Oct. 1981. Mém. Mus. Nat. Hist. nat. Paris, Sér. C*, **49**, 43–48.

Conard-Noireau, M. (1983). La dynamique des dépôts cénomaniens de Haute-Provence: observations nouvelles et implications paléogéographiques. *Bull. Soc. géol. Fr. Sér. 7.* **25** (2), 239–246.

Conard-Noireau, M., Busnardo, R., Ducreaux, J.L., Gaillard, M.G., Philip, J., Babinot, J.F., Tronchetti, G., Bilotte, M., Alonso-Millan, A., Amiot, M., Floquet, M., Mathey, B., Deloffre, R., Platel, J.P., Louail, J., Viaud, J.M., Monciardini, C., Alcayde, G., Robaszynski, F., Magniez, F., Rat, P. and Charollais, J. (1982). Conclusions au colloque sur le Turonien II. Cartes des facies pour le Turonien de la France et des régions adjacentes. In: *Proc. Colloque sur le Turonien, Paris, Oct. 1981, Mém. Mus. Nat. Hist. nat. Paris, Sér. C.* **49**, 231–241.

Conklin, J. and Moore, C. (1977). Paleoenvironmental analysis of the Lower Cretaceous Cupido Formation, northeast Mexico. In:

Bebout, D.G. and Loucks, R.G. (eds.) *Cretaceous carbonates of Texas and Mexico: applications to subsurface exploration, Rep. invest. Bur. econ. Geol. univ. Texas.* **89**, 302–323.

Contescu, L.R. (1974). Geologic history and paleogeography of eastern Carpathians: example of Alpine geosyncline evolution. *Bull. Am. Assoc. Petrol. Geol.* **58** (12), 2436–2476.

Contini, D. (1969). L'Aalenien et le Bajocien du Jura Franc-Comtois. Etude stratigraphique. *Ann. sci. univ. Besancon, 3rd Sér., Géol.* **11**, 204pp.

Contini, D. (1979). Relations entre les bassins sédimentaires Souabo-Lorrain et Jurassico-Dauphinois au Dogger. Naissance, évolution et disparition d'une plate-forme carbonatée. In: *La sédimentation Jurassique W. européen, Proc. Symp. Paris, May 1977, Assoc. Sédimentol. Fr. Spéc. Publ.* **1**, 125–134.

Contini, D. and Lamaud, P. (1978). Paléogéographie et paléoécologie du Toarcien inférieur du jura et de la bordure méridionale des Vosges. *Ann. sci. Univ. Besancon, 3rd Sér., Géol.* **29**, 21–36.

Cook, T.D. and Bally, A.W. (eds.) (1975). *Stratigraphic Atlas of North and Central America.* (Explor. Dept. Shell Oil Co., Houston, Texas) Princeton Univ. Press, Princeton, 272pp.

Corroy, G. (1963). L'évolution paléogéographique post-hercynienne de la Provence. In: Durand-Delga, M. (ed.) *Livre à la Mémoire du Professeur Paul Fallot: L'évolution paléogéographique et structurale des domaines méditerranéens et Alpins d'Europe. Mém. Soc. géol. Fr. Hors Sér. 1,* **2**, 19–44.

Corsin, P. and Stamppfli, G. (1977). La Formation de Shemshak dans l'Elburz oriental (Iran), flore-stratigraphie-paléogéographie. Géobios. **10** (4), 509–571.

Corvalán, J. (1979). Rasgos geológico-estructurales y metalogenéticos relacionados con la segmentación de los Andes. In: *Acta 2nd Congr. geol. Chileno, Arica, Chile Aug. 1979, 4, Inst. Investig. geol. Ciudad de Arica.* 45–75.

Corvalán, J. and Vergara, M. (1980). Presencia de fósiles marinos en las calizas de Polpaico. Implicaciones paleoecológicas y paleogeográficas. *Rev. geol. Chile.* **10**, 75–83.

Costea, I., Cosma, D. and Vinogradov, C. (1978). Microfacies urile Cretaciculi Inferior din Platforma Moesic. *Studi Cercetari. Geol. Geofiz. Geogr.* **23** (2), 299–311.

Costea, I., Vinogradov, C., Cosma, D. and Bonig, H. (1981). Studiul mocrofacial al depozitelor Cretacice din Platforma Moldoveneasca si Depresunea Birladului. *Studi. Cercetari. Geol. Geofiz. Geogr.* **26** (2), 275-292.

Cotillon, P. (1966). Stratigraphie et paléogéographie du Crétacé Inférieur de l'anticlina de Gourdan et du bassin du Var Moyen de Daluis à la Mescla. *Bull. Soc. géol. Fr. Sér. 7.* **8** (3), 362–375.

Cotillon, P. (1971). Le Crétacé Inférieur de l'arc subalpin de Castellane entre l'Asse et le Var: stratigraphie et sédimentologie. *Mém. Bur. Rech. géol. minières.* **68**, 313pp.

Cotillon, P. (1974). Sédimentation rythmique et milieux de dépôt: données fournies par l'étude du Crétacé Inférieur de l'arc subalpin de Castellane (France SE). *Bull. Soc. géol. Fr. Sér. 7.* **16** (6), 583–592.

Cotillon, P., Ferry, S., Busnardo, R., Lafarge, D. and Renaud, B. (1979). Synthèse stratigraphique et paléogéographique sur les facies urgoniens du Sud de l'Ardèche et du nord du Gard (France). In: *L'Urgonien des Pays Méditerranéens. Géobios Spéc. Mém,* **3**, 121–140.

Cotillon, P., and Medus, J. (1971). Données stratigraphiques et palynologiques sur un horizon Albien dans l'arc de Castellane

(Basses-Alpes, Alpes-Maritimes, Var): conséquences paléogéographiques. *Géobios.* **4** (2), 101–116.

Courel, L. (1962). Les facies à évaporites rapportées au Trias sur la bordure nord-est du Massif central. *Bull. Soc. géol. Fr. Sér. 7.* **4**, suppl., 20–22.

Courel, L. (1973). Modalités de la transgression Mésozoïque Trias et Rhétien de la bordure nord – est du Massif Central Francais. *Mém. Soc. géol. Fr. Hors Sér.* **118**, 152pp.

Courel, L. (1982). Les facies de transgression du Trias sur la bordure Nord-est du Massif Central Francais. In: *Proc. Symp. Triassic global geology, Würzburg, Feb. 1982, Geol. Rdsch.* **71** (3), 773–781.

Courel, L., Durand, M., Gall, J-C. and Juraine, G. (1973). Quelques aspects de la transgression Triassique dans le nord-est de la France. Influence d'un Eperon Bourguignon. *Rev. Géogr. phys. Géol. dyn.* **15** (5), 547–554.

Courel, L., Feuillee, P., Rat, P., Sedoh, F. and Trescartes, J. (1972). Les sables Albiens dans le sud-est du bassin Parisien: analyse sédimentologique. Essai paléogéographique. *Rev. Géogr. phys. Géol. dyn.* **14** (2), 171–188.

Cousin, M. (1970). Esquisse géologique des confins italo-yougoslaves: leur place dans les Dinarides et les Alpes méridionales. *Bull. Soc. géol. Fr. Sér. 7.* **122** (6), 1034–1047.

Cousin, M., Delga, M.D. and Sigal, J. (1966). Données nouvelles sur le Crétacé de la haute et moyenne vallée du Tagliamento (Alpes méridionales, Frioul, Italie). *Bull. Soc. géol. Fr. Sér. 7.* **8** (4), 503–517.

Cousin, M. and Fourcade, E. (1982). Les facies Crétacés de la plate-forme Frioulano-Karstique et leur environnement paléogéographique (Alpes méridionales et Dinarides). In: Cotillon, P. and Phillip, J. (eds.) *Proc. Mtg. Les facies de Plate-forme du Crétacé dans le domaine Mésogéen, 26th Int. geol. Congr., Paris, 1980, Cret. Res.* **3** (1–2), 113–124.

Crescenti, U. (1971). Biostratigraphic correlations in the Jurassic facies of central Italy by means of microfossils. In: Vegh-Neubrandt, E. (ed.) *Colloque du Jurassique Méditerranéen, Budapest, 1969, Ann. Inst. geol. publici Hung.* **54** (2), 209–213.

Cros, P. and Fryssalakis, G. (1982). Un exemple de sédimentation volcanogénique en bassin pélagique: Trias Moyen des Alpes Camiques orientales. In: Lang, J. (ed.) *Livre Jubilaire Gabriel Lucas: géologie sédimentaire. Mém. Géol. Univ. Dijon.* **7**, 139–152.

Crostella, A. and Barter, T. (1980). Triassic-Jurassic depositional history of the Dampier and Beagle sub-basins, northwest shelf of Australia. *Aust. Petrol. Explor. Assoc. J.* **20** (1), 25–33.

Crouch, J.K. (1981). Northwest margin of California continental borderland: marine geology and tectonic evolution. *Bull. Am. Assoc. Petrol. Geol.* **65** (12), 191–218.

Csaszar, G. and Haas, J. (1979). Review of the Facies and palaeogeography of the Cretaceous in Hungary. In: Wiedmann, J. (ed.) *Aspekte der Kreide Europas, proc. 1st Symp. German Cretaceous, Munster, April 1978, Int. Union geol. Sci., Ser. A.* **6**, E. Schweizerbart'sche Verlagsbuchhandlung, Stuttgart, 413–424.

Cugny, P. and Rey, J. (1975). Un exemple d'utilisation de l'analyse factorielle des correspondances en paléoécologie: la répartition des microfossiles dans le Bedoulien d'Estremadura (Portugal). *Bull. Soc. géol. Fr. Sér. 7.* **17** (5), 787–796.

Cumella, S.P. (1983). Relation of Upper Cretaceous regressive sandstone units of the San Juan Basin to source area tectonics. In: Reynolds, M.W. and Dolly, E.D. (eds.) *Mesozoic Paleogeography of the*

West-Central United States. Proc. 2nd Rocky Mt. Paleogeogr. Symp., Denver, Colorado. Rocky Mt. Sect. Soc. econ. Paleontol. Mineral. 189–200.

Curnelle, R. (1983). Evolution structuro-sédimentaire du Trias et de l'infra-Lias d'Aquitaine. *Bull. Cent. Rech. Explor. Prod. Elf-Aquitaine.* 7 (1), 69–99.

Curnelle, R., Dubois, P. and Seguin, J-C. (1980). Le Bassin d'Aquitaine: substratum ante-tertiare et bordures mésozoïques. In: Lorenz, C. (ed.) *Geologie des Pays européens: France, Belgique, Luxembourg. 26th Int. Geol. Congr. Paris.* 231–242.

Curnelle, R., Dubois, P. and Seguin, J-C. (1982). The Mesozoic-Tertiary evolution of the Aquitaine Basin. In: Kent, P., Bott, M.H.P., McKenzie, D.P. and Williams, C.A. (eds.) *The evolution of sedimentary basins. Proc. Symp. R. Soc. June 1981. Philos. Trans. R. Soc. Lond.* A305 (1489), 63–84.

Curray, J.R. and Moore, D.G. (1974). Sedimentary and tectonic processes in the Bengal deep-sea fan and geosyncline. In: Burk, C.A. and Drake, C.L. (eds.) *The geology of continental margins. Springer-Verlag, New York,* 617–628.

Czerminski, J. and Pajchlowa, M. (eds.) (1975). *Lithologic-palaeogeographic atlas of the platform regions in Poland, 1:2,000,000, II, the Mesozoic (exclusive of the Upper Cretaceous).* Wydawn Geol., Warsaw, 13 sheets. (Not seen, unobtainable in U.K. even through I.L.L. z search).

Dadlez, J. (1976). Żelaziste osady Oksfordu w Niece Pomorskiej. *Biul. Inst. geol.* 295, 137–167.

Dadlez, R. (1969). Stratygrafia liasu w Polsce Zachodneij. *Pr. Inst. geol. Warszawa.* 57, 92pp.

Dadlez, R. (1976). Perm i mezozoik nieki Pomorskiej. *Pr. Inst. geol. Warszawa.* 79, 173pp.

Dadlez, R. and Kopik, J. (1975). Stratigraphy and palaeogeography of the Jurassic. In: Slowanska, B. and Wilczek, M. (eds.) *Proc. 5th Anniv. Symp. geol. Inst., Warsaw, June 1970, Bull. geol. Inst. Warsaw.* 252, 149–171.

Dahm, H. (1965). Stratigraphie und Palaeogeographie im Kantabrischen Jura (Spanien). In: Beuther, A., Dahm, H., Kneuper-Haack, F. *et al.* (1966). *Der Jura und Wealden in Nordost-Spanien. Beih. geol. Jahrb.* 44, 13–54.

Dalland, A. (1973). Mesozoic rocks of Andoya, northern Norway. In: Whiteman, A., Roberts, D. and Sellevoll, M.A. (eds.) *Petroleum geology and the geology of the North Sea and northeast Atlantic continental margin, Proc. Conf. Bergen, 1973, Nor. geol. Unders.* 316, 271–288.

Dalland, A. (1979). The sedimentary sequence of Andoy, northern Norway – depositional and structural history. *Proc. Norwegian North Sea Symp., Tromso, Norway, 1979, Nor. Petrol. Foren.* NNSS/26 31pp.

Dalland, A. (1981). Mesozoic sedimentary succession at Andoy, northern Norway, and relation to structural development of the north Atlantic area. In: Kerr, J.W. and Fergusson, A.J. (eds.) *Geology of the North Atlantic borderlands, Proc. Symp. St. John's, Newfoundland, June 1979, Mem. Can. Soc. Petrol. Geol.* 7, 563–584. (NB. Virtually identical to Dalland, A. 1979.)

Dalmayrac, B. (1978). Géologie des Andes Péruviennes. Géologie de la Cordillère orientale de la région de Huanuco: sa place dans une transversale des Andes du Pérou central (9° 5' a 10° 30' S). *Trav. Doc. O.R.S.T.O.M.* 93, 161pp.

Dalmayrac, B., Laubacher, G. and Marocco, R. (1980). Géologie des Andes Péruviennes: caractères généraux de l'évolution géologique des Andes péruviennes. *Trav. Doc. O.R.S.T.O.M.* 122, 501pp.

Damborenea, S.E. and Mancenido, M.O. (1979). On the palaeogeographical distribution of the pectinid genus *Weyla* (Bivalvia, Lower Jurassic). *Palaeogeogr. Palaeoclimatol. Palaeoecol.* 27 (1–2), 85–102.

Damotte, R. and Fourcade, E. (1971). *Neocyprideis murciensis* n.sp., ostracode nouveau du Maestrichtien de la province de Murcie (Sud-Est de l'Espagne). *Bull. Soc. géol. Fr. Sér. 7.* 13 (1–2), 169–173.

Dane, C.H. (1960). New information on the areal extent of some Upper Cretaceous units in northwestern New Mexico. *Prof. pap. U.S. geol. Surv.* 400-B, 241–243.

Dane, C.H., Wanek, A.A. and Reeside, J.B. (1957). Reinterpretation of section of Cretaceous rocks in Alamosa Creek Valley area, Catron and Socorro counties, New Mexico. *Bull. Am. Assoc. Petrol. Geol.* 41 (2), 181–196.

Dardeau, G. and Marchand, D. (1981). Présence d'*Amoeboceras newbridgense* Sykes & Callomon, ammonite boréale, dans une fauna submésogéenne de l'Oxfordien Supérieur (zone à bifurcatus) du sud est de la France (Alpes Maritimes): intérêt biostratigraphique, paléogéographique et paléobiogéographique. *Géobios.* 14 (3), 407–413.

Dardeau, G., Marchand, D. and Thierry, J. (1979). Relations entre la répartition des facies et la composition des faunes d'Ammonites au cours de l'évolution du bassin Callovien des Alpes-Maritimes. *Bull. Soc. géol. Fr. Sér. 7.* 21 (6), 753–757.

Dardel, R.A. and Rosset, R. (1971). Histoire géologique et structurale du bassin de Parentis et de son prolongement en mer. In: *Proc. Symp. Histoire Structurale du Golfe de Gascogne, Rueil-Malmaison, Déc. 1970, Publ.Inst. Fr. Pet., collection colloq. et Séminaires.* 22 (1), Editions Technip, Paris, IV-2, 1–28.

Davies, D.K. (1969). Shelf sedimentation: an example from the Jurassic of Britain. *J. sediment. Petrol.* 39 (4), 1344–1370.

Davis, G.A., Monger, J.W.H. and Burchfiel, B.C. (1978). Mesozoic construction of the Cordilleran 'collage', central British Columbia to central California. In: Howell, D.G. and McDougall, K.A. (eds.) *Mesozoic paleogeography of the western United States, Proc. 2nd Symp. Pacific Coast Paleogeogr, April 1978, Pacific Sect. Soc. econ. Paleontol. Mineral.* 1–32.

Davis, J.C. (1970). Petrology of Cretaceous Mowry Shale of Wyoming. *Bull. Am. Assoc. Petrol. Geol.* 54 (3), 487–502.

Davoudzadeh, M. and Schmidt, K. (1981). Contribution to the paleogeography and stratigraphy of the Upper Triassic to Middle Jurassic of Iran. *Neues Jahrb. Geol. Palaontol. Abhandlungen.* 162 (2), 137-163.

Davoudzadeh, M. and Schmidt, K. (1982). Zur Trias des Iran. In: *Proc. Symp. Triassic global geology, Würzburg, Feb. 1982, Geol. Rdsch.* 71 (3), 1021-1039.

Davoudzaden, M. and Schmidt, K. (1983). Contribution to the paleogeography, stratigraphy and tectonics of the Middle and Upper Jurassic of Iran. *Neues Jahrb. Geol. Palaontol. Abhandlungen.* 166 (3), 327-346.

Dayczak-Calikowska, K. (1966). Rozprzestrzenienie osadów najniższego keloweju na Niżu Polskim. *Kwart. geol.* 10 (1), 74–87.

Dayczak-Calikowska, K. (1977). Baton górny i kelowej w Potnocno-Zachodniej Polsce. *Pr. Inst. geol. Warszawa.* 84, 69pp.

Debrand-Passard, S. (1982). Le Jurassique Supérieur du Berry (sud du Bassin de Paris). *Mém. Bur. Rech. géol. minières.* 119, 227pp.

Debrand-Passard, S., Lorenz, J. and Tintant, H. (1974). Précisions sur le passage Dogger-Malm et la série Jurassique Supérieur dans le sud du bassin de Paris (région d'Issoudun, Indre). *Bull. Soc. géol. Fr. Sér. 7.* **16 (4)**, 470–475.

Debrand-Passard, S., Marchand, D., Lorenz, J. and Odin, G-S. (1980). Les variations de facies dans le département du Cher à la limite Dogger-Malm. Tentative de reconstitution paléogéographique. *Bull. Soc. géol. Fr. Sér. 7.* **22** (4), 567–572.

De Cserna, Z. (1960). Orogenesis in time and space in Mexico. *Geol. Rdsch.* **50**, 595–605.

De Cserna, Z. (1970). Mesozoic sedimentation, magmatic activity and deformation in Northern Mexico. In: Seewald, K.O. and Sundeen, D. (eds.) *The geologic framework of the Chichuahua tectonic belt, a symposium in honor of Professor Ronald K. De Ford, Midland, Texas Nov. 1970, West Texas Geol. Soc.* **71–59**, 99–117.

De Cserna, Z., Palacio-Nieto, M. and Pantoja-Alor, J. (1978). Relaciones de facies de las rocas Cretácicas en el noroeste de Guerrero y en areas colindantes de México y Michoacán. *Rev. Inst. Geol. Univ. nac. Autónoma México.* **2** (2), 8–18.

De Ford, R.K. and Haenggi, W.T. (1970). Stratigraphic nomenclature of Cretaceous rocks in northeastern Chihuahua. In: Seewald, K.O. and Sundeen, D. (eds.) *The geologic framework of the Chihuahua tectonic belt, a symposium in honor of Professor Ronald K. De Ford, Midland, Texas, Nov. 1970, Publ. West Texas Geol. Soc.* **71–59**, 175–196.

De Lapparent, A.F., Philip, J., Marie, P. and Dufaure, P. (1974). Le Montagne de Parandaz et les transgressions Crétacées en Afghanistan central. *Bull. Soc. géol., Fr. Sér. 7.* **16** (3), 311–318.

Delfaud, J. (1971). Réflexions sur l'environnement Jurassique. L'exemple des séries Périgord-Quercy. *Rev. Géogr. phys. Géol. dyn.* **13** (3), 214–231.

Delfaud, J. (1974). La sédimentation deltaïque ancienne – exemples nord-Sahariens. *Bull. Cent. Rech. Pau.* **8** (1), 241–262.

Delfaud, J. (1980). Le contexte paléogéographique des séquences d'échelle Moyenne dans le Jurassique aquitain. *Bull. Soc. géol. Fr. Sér. 7.* **22** (4), 573–579.

Delfaud, J. (1983). Les paléoclimats du Jurassique en Europe occidentale. *Bull. Inst. géol. Bassin d'Aquitaine.* **34**, 121–35.

Deloffre, R. (1978). Histoire géologique du Crétacé Moyen en Aquitaine (France S.W.). In: Philip, J. (ed.) *Colloque sur le Cénomanien: France-Europe occidentale. Paris, Sept. 1976, Géol. Méditerr., Univ. Provence.* **5** (1), 79–86.

Del Pozo, J.R. and Tomas, M.J.A. (1972). Consideraciones sedimentológicas y paleogeográficas de las facies Purbeck y Weald de la cubeta de Santander-Burgos. *Estud. geol. Inst. Invest. geol. Lucas Mallada.* **28**, 173–192.

Dembowska, J. (1973). Portland na nizu Polski. *Pr. Inst. geol. Warszawa.* **70**, 107pp.

Denham, J.I. and Brown, B.R. (1976). A new look at the Otway Basin. *Aust. Petrol. Explor. Assoc. J.* **16** (1), 91–98.

Dercourt, J. (1962). Contribution à l'étude géologique du Péloponnèse. Terminaison paléogéographique du haut-fond du Parnasse. *Bull. Soc. géol. Fr. Sér. 7.* **4** (3), 340–356.

Deshpande, S.V. and Merh, S.S. (1980). Mesozoic sedimentary model of Wagad Hills, Kutch, Western India. *J. geol. Soc. India.* **21** (2), 75–83.

Desio, A. (ed.) (1973). *Geologia dell'Italia. Unione Tipografica-Editrice Torinese,* 108lpp.

Dessanti, R.N. (1958). Geologic outline of natural regions of Men-

doza, Argentina. *Bull. Am. Assoc. Petrol. Geol.* **42** (11), 2670–2691.

Detterman, R., (1973). Mesozoic sequence in Arctic Alaska. In: Pitcher, M.G. (ed.) *Arctic Geology, Proc. 2nd Int. Symp., San Francisco, February 1971, Mem. Am. Assoc. Petrol. Geol.* **19**, 376–387.

Dickins, J.M. (1973). The geological sequence and the Permian-Triassic boundary in Australia and eastern New Guinea. In: Logan, A. and Hills, L.V. (eds.) *The Permian and Triassic systems and their mutual boundary, Proc. Symp. Univ. Calgary, Aug. 1971. Mem. Can. Soc. Petrol. Geol.* **2**, 425–432.

Dickinson, W.R. (1976). Sedimentary basins developed during evolution of Mesozoic-Cenozoic arc-trench system in western North America. *Can. J. Earth Sci.* **13** (9), 1268–1287.

Dickinson, W.R., Beard, L.S., Brackenridge, G.R., Erjavec, J.L., Ferguson, R.C., Inman, K.F., Knepp, R.A., Lindberg, F.A. and Ryberg, P.T. (1983). Provenance of North American phanerozoic sandstones in relation to tectonic setting. *Bull. geol. Soc. Am.* **94** (2), 222–235.

Diener, I. (1967a). Die Paläogeographie der Kreide im Nordteil der DDR in Beziehung zu den Nachbargebieten. *Ber. Deutsch. Ges. geol. Wiss., Ser. A, Geol. Paläeontol.* **12** (3–4), 289–313.

Diener, I. (1967b). Zur Paläogeographie der Oberkreide Nordostdeutschlands. *Ber. Deutsch. Ges. geol. Wiss., Ser. A, Geol. Paläeontol.* **12** (5), 493–509.

Diener, I. (1968). Kreide. In: Schmidt, K. (ed.) *Grundriss der Geologie der Deutschen Demokratischen Republik,* **1**, *Geologische Entwicklung des Gesamtgebietes.* Akademie Verlag Berlin, 320–342 (+ Anl.-Abb. 17–20).

Dingle, R.V. (1973). Mesozoic palaeogeography of the southern Cape, South Africa. *Palaeogeogr. Palaeoclimatol. Palaeoecol.* **13** (3), 203–213.

Dingle, R.V. (1978). South Africa. In: Moullade, M. and Nairn, A.E.M. (eds.) *The Phanerozoic geology of the world,* **2**, *The Mesozoic, A.* Elsevier, Amsterdam, 401–434.

Dingle, R.V. (1982). Some aspects of Cretaceous ostracod biostratigraphy of South Africa and relationships with other Gondwanide localities. *Cret. Res.* **3** (4), 367–389.

Dingle, R.V., Siesser, W.G. and Newton, A.R. (1983). *Mesozoic and Tertiary geology of Southern Africa.* A.A. Balkema, Rotterdam, 375pp.

Di-Nocera, S. and Scandone, P. (1977). Triassic nanoplankton limestones of deep basin origin in the central Mediterranean region. *Palaeogeogr. Palaeoclimatol. Palaeoecol.* **21** (2), 101–111.

Dixon, J. (1979). The Lower Cretaceous Atkinson Point Formation (new name) on the Tuktoyaktuk Peninsula, N.W.T.: a coastal fan-delta to marine sequence. *Bull. Can. Pet. Geol.* **27** (2), 163-182.

Dixon, J. (1982a). Jurassic and Lower Cretaceous subsurface stratigraphy of the Mackenzie Delta – Tuktoyaktuk Peninsula, N.W.T. *Bull. geol. Surv. Can.* **349**, 52pp.

Dixon, J. (1982b). Upper Oxfordian to Albian geology, Mackenzie Delta, Arctic Canada. In: Embry, A.F. and Balkwill, H.R. (eds.) *Arctic geology and geophysics, Proc. 3rd Int. Symp. Arctic geol., Calgary, June 1981, Mem. Can. Soc. Pet. geol.* **8**, 29–42.

Dommergues, J-L. (1982). Le provincialism des ammonites nord-ouest européennes au Lias moyen. Une crise faunique sous contrôle paléogéographique. In: *Proc. Symp. Biogéographie et tectonique des plaques, Brest, Mar. 1982. Bull. Soc. géol. Fr. Sér. 7.* **24** (5–6), 1047–1051.

Donoso, J.M.G. and López-Garrido, A.C. (1975). Características bioestratigráficas y paleogeográficas del Cretácico Superior de la Sierra del Segura. (N.E. de la Provincia de Jaen, Zona Prebética).

In: *Primer coloquio de estratigrafía y paleogeografía del Cretácico de España. Bellaterra-Tremp, November 1973, Publ. Empresa Nac. Adaro. Invest. Mineras, Ser. 7, Trabajos de congresos y reuniones, Madrid.* 1, 45–62.

Donovan, D.T. (1967). The geographical distribution of Lower Jurassic ammonites in Europe and adjacent areas. In: Adams, C.G. and Ager, D.V. (eds.) *Aspects of Tethyan biogeography.* Systematics Assoc. Publ. 7, 111–134.

Donovan, D.T., Horton, A. and Ivimey-Cook, H.C. (1979). The transgression of the Lower Lias over the northern flank of the London Platform. *J. geol. Soc. Lond.* 136 (2), 165–173.

Dott, R.H., Jnr., Winn, R.D., Jnr. and Smith C.H.L. (1982). Relationship of Late Mesozoic and Early Cenozoic sedimentation to the tectonic evolution of the southernmost Andes and Scotia Arc. In: Craddock, C. (ed.) *Antarctic geoscience. Proc. Symp. Antarctic geology and geophysics, Madison, Wisconsin Aug. 1977, Int. Union geol. Sci. Ser. B.* 4, Univ. Wisconsin Press, 193–202.

Douglas, R.J.W., Gabrielse, H., Wheeler, J.O., Stott, D.F. and Belyea, H.R. (1970). Geology of western Canada. In: Douglas, R.J.W. (ed.) *Geology and economic minerals of Canada, Econ. Geol. Rep. geol. Surv. Can.* 1, 365–488.

Drittenbass, W. (1979). Sedimentologie und Geochemie von Eisen-Mangan führenden Knollen und Krusten im Jura der Trento-Zone (östliche Südalpen, Norditalien). *Eclog. geol. Helv.* 72 (2), 313-345.

Druckman, Y. (1974a). The stratigraphy of the Triassic sequence in southern Israel. *Bull. geol. Surv. Israel.* 64, 92pp.

Druckman, Y. (1974b). Triassic Palaeogeography of Southern Israel and the Sinai Peninsula. In: Zapfe, J. (ed.) *The stratigraphy of the Alpine-Mediterranean Triassic, Proc. Symp. Wien, May 1973.* Springer-Verlag, Wien, 79–86.

Druckman, Y., Hirsch, F. and Weissbrod, T. (1982). The Triassic of the southern margin of the Tethys in the Levant and its correlation across the Jordan rift valley. In: *Proc. Symp. Triassic global geology, Wurzburg, Feb. 1982, Geol. Rdsch.* 71 (3), 919–936.

Drummond, P.V.O. (1970). The mid-Dorset swell. Evidence of Albian-Cenomanian movements in Wessex. *Proc. Geol. Assoc. Lond.* 81 (4), 679–714.

Dubar, G. (1963). Notes sur la paléogéographie du Lias Marocain (domaine atlasique). In: Durand-Delga, M. (ed). *Livre à la Mémoire du Professeur Paul Fallot; L'évolution paléogéographique et structurale des domaines méditerranéens et Alpins d'Europe. Mém. Soc. géol. Fr. Hors Sér.* 1, 1, 529–544.

Dubar, G. and Mouterde, R. (1978). L'Aalenien et le Toarcien terminal du Haut-Atlas. Esquisse paléogéographique. *Bull. Soc. géol. Fr. Sér. 7.* 20 (2), 169–178.

Dubois, P. and Seguin, J-C. (1978). Les flyschs Crétacé et Eocène de la zone commingeoise et leur environnement. *Bull. Soc. géol. Fr. Sér. 7.* 20 (5), 657–671.

Dubois, P. and Umbach, P. (1974). A propos du Trias de deux bassins sédimentaires francais: le Bassin de Paris et le Bassin du Sud-Est. *Bull. Soc. géol. Fr. Sér. 7.* 16 (6), 696–707.

Duchrow, H. (1968). Stratigraphie und Lithologie des Keupers im Lippischen Berglande. *Z. Deutsch. geol. Ges.* 117 (2-3), 371-387.

Du Dresnay, R. (1971). Extension et développement des phénomènes récifaux Jurassiques dans le domaine atlasique marocain, particulièrement au Lias Moyen. *Bull. Soc. géol. Fr. Sér.* 13 (1–2), 46–56.

Duffaud, F. (1960). Contribution à l'étude stratigraphique du bassin

secondaire du Haut-Atlas occidental (Sud-Ouest Marocain). *Bull. Soc. géol. Fr. Sér. 7.* 2 (6), 728–734.

Dumont, J-F., Uysal, S. and Monod, O. (1980). La série de Zindan: un élément de liaison entre plate-forme et bassin à l'Est d'Isparta (Taurides occidentales, Turquie). *Bull. Soc. géol. Fr. Sér. 7.* 22 (2), 225–232.

Duque-Caro, H. (1980). Geotectónica y evolución de la región noroccidental Columbiana. *Bol. geol. Inst. Nac. Invest. Geol. Min (INGEOMINAS).* 23 (3), 4–37.

Edlinger, G. (1966). Zur Geologie des Weissen Jura zwischen Solnhofen und Eichstatt (Mfr.). *Erlanger geol. Abhandlungen.* 612, 20pp.

Edwards, C.W. (1982). New paleontologic evidence of Triassic sedimentation in West Antarctica. In: Craddock, C. (ed.) *Antarctic geoscience. Proc. Symp. Antarctic geology and geophysics, Madison, Wisconsin Aug. 1977, Int. Union geol. Sci. Ser. B.* 4, Univ. Wisconsin Press, 325–330.

Eebum, C. (1976). Geotektonische und paläogeographische Bedingungen der mesozoischen Sedimentgenese und wichtigste Gesetzmässigkeiten der Kohlebildung in der Mongolischen Volksrepublik. *Jahrb. Geol.* 7–8, (1971–1972), 325–336.

Eicher, D.L. (1966). Depth in the Greenhorn Sea. In: Kauffmann, E.G. and Kent, H.C. (eds.) *Paleoenvironments of the Cretaceous seaway in the Western Interior, Prepr. Symp. Colorado School of Mines, May 1967, Dept. Publ. Colorado School of Mines, Golden, Colorado.* 145–172.

Eicher, D.L. (1969). Paleobathymetry of Cretaceous Greenhorn sea in eastern Colorado. *Bull. Am. Assoc. Petrol. Geol.* 53 (5), 1075–1090.

Einsele, G. and Wiedmann, J. (1982). Turonian black shales in the Moroccan coastal basins: first upwelling in the Atlantic Ocean? In: Rad, U. von, Hinz, K., Sarnthein, M. and Seibold, E. (eds.) *Geology of the Northwest African continental margin.* Springer-Verlag, 396–414.

Eisbacher, G.H. (1977). Mesozoic-Tertiary basin models for the Canadian Cordillera and their geological constraints. *Can. J. Earth Sci.* 14 (10), 2414–2421.

Eisbacher, G.H. (1981). Late Mesozoic-Paleogene Bowser Basin molasse and cordilleran tectonics, western Canada. In: Miall, A.D. (ed.) *Sedimentastion and tectonics in alluvial basins, Spec. Pap. geol. Assoc. Can.* 23, 125–152.

Elliott, G.F. (1977). Inferred isocrymal distribution of Jurassic dasycladacean Algae in Europe, north Africa and southwestern Asia. *J. geol. Soc. Lond.* 133 (4), 363–373.

Elliott, J.L. (1972). Continental drift and basin development in south eastern Australia. *Aust. Petrol. Explor. Assoc. J.* 12 (2), 46–51.

Elmi, S., Almeras, Y., Ameur, M.H., Atrops, F., Benhamon, M. and Moulan, G. (1982). La dislocation des plates-formes carbonatées liassiques en Méditerranée occidentale et ses implications sur les échanges fauniques. In: *Proc. Symp. Biogéographie et tectonique des plaques, Brest, Mar. 1982. Bull. Soc. géol. Fr. Sér. 7.* 24 (5–6), 1007–1016.

El Shazly, E.M. (1977). The geology of the Egyptian region. In: Nairn, A.E.M., Kanes, W.H. and Stehli, F.G. (eds.) *The ocean basins and margins,* 4A, *The eastern Mediterranean.* Plenum Press, London, 379–444.

Embray, A.F. (1982). The Upper Triassic-Lower Jurassic Heiberg deltaic complex of the Sverdrup Basin. In: Embry, A.F. and Balkwill, H.R. (eds.) *Arctic geology and geophysics, Proc. 3rd Int. Symp. Arctic Geol., Calgary, June 1981, Mem. Can. Soc. Pet. geol.* 8, 189–217.

Emmert, U. (1981). Trias nördlich der Alpen, 3.2, Muschelkalk. In:

Haunschild, H. and Jerz, H. (eds.) *Erläuterungen zur Geologischen Karte von Bayern 1:500,000.* Bayerisches Geologisches Landesamt, München, 46–49.

Enay, M.R. (1965). Les formations coralliennes de Saint-Germain-de-Joux (Ain). *Bull. Soc. géol. Fr. Sér. 7.* **7** (1), 23–31.

Enay, R. (1980). Indices d'émersion et d'influences continentales dans l'Oxfordien supérieur-Kimmeridgien inférieur en France. Interprétation paléogéographique et conséquences paléobiogéographiques. *Bull. Soc. géol. Fr. Sér. 7.* **22** (4), 581–590.

Enay, R. (1980). Paléobiogéographie et Ammonites jurassiques: 'rythmes fauniques' et variations du niveau marin; voies d'échanges, migrations et domaines biogéographiques. In: Livre Jubilaire du cent cinquantenaire 1830-1980. *Mém. Soc. géol. Fr. Hors Sér.* **10**, 261–281.

Enay, R. and Mangold, C. (1982). Dynamique biogéographie et évolution des faunes d'ammonites au Jurassique. In: *Proc. Symp. Biogéographie et tectonique des plaques, Brest, Mar. 1982. Bull. Soc. géol. Fr. Sér. 7.* **24** (5–6), 1025–1046.

Enay, R., Mangold, C., Cariou, E., Contini, D., Debrand-Passard, S., Donze, P., Gabilly, J., Lafavrais-Raymond, A., Mouterde, R. and Thierry, J. (Le Groupe Francais d'Etude du Jurassique) Synthèse paléogéographique du Jurassique francais. *Doc. Lab. Géol. Fac. Sci. Lyon, Hors Sér.* **5**, 210pp.

Enos, P. (1974). Reefs, platforms and basins of Middle Cretaceous in northeast Mexico. *Bull. Am. Assoc. Petrol. Geol.* **58** (5), 800–809.

Enos, P. (1977). Tamabra Limestone of the Poza Rica Trend, Cretaceous, Mexico. In: Cook, H.E. and Enos, P. (eds.) *Deep water carbonate environments. Spec. Publ. Soc. econ. Paleontol. Mineral.* **25**, 273–314.

Enos, P. (1983). Late Mesozoic paleogeography of Mexico. In: Reynolds, M.W. and Dolly, E.D. (eds.) *Mesozoic Paleogeography of the West-Central United States. Proc. 2nd Rocky Mt. Paleogeogr. Symp., Denver, Colorado. Rocky Mt. Sect. Soc. econ. Paleontol. Mineral.* 133–158.

Erben, H.K. (1956). El Jurásico Medio y el Calloviano de México. *Monograf. 20th Int. geol. Congr. México.* 140pp.

Erben, H.K. (1957). Paleogeographic reconstructions for the Lower and Middle Jurassic and the Callovian of Mexico. In: *Proc. 20th Int. geol. Congr., Ciudad de Mexico 1956, Sect. II, El Mesozoico del hemisferio occidental y sus correlaciones mundiales.* 35–43.

Erlangen, A.Z. (1977). Jurassic stratigraphy of Franconia. *Stutt. Beitr. Naturkd. Ser. B.* **31**, 32pp.

Ernst, G., Schmid, F. and Seibertz, E. (1983). Event stratigraphie im Cenoman und Turon von N.W. Deutschland. In: *2nd Symp. Kreide, München, 1982. Zitteliana.* **10**, 531–54.

Escobar, E.A. (1977). Acerca de la paleogeografía Neocomiana en la region al sur de Copiapo. (Provincia de Atacama -Chile). *Rev. Assoc. geol. Argentina.* **32** (1), 24–33.

Esquevin, J., Fournie, D. and De Lestang, J. (1971). Les séries de l'Aptien et de l'Albien des régions nord-Pyrénéennes et du sud Aquitain. (France-Sud). *Bull. Cent. Rech. Pau.* **5** (1), 87–151.

Etayo-Serna, F., Renzoni, G. and Barrero, D. (1976). Contornos sucesivos del mar Cretaceo en Colombia. In: Etayo-Serna, F. and Caceres-Giron, C. (eds.) *Mem. 1st Congr. Colombia Geol., Univ. nac. Colombia, Bogota, August 1969, Bogota,* 217–252.

Evans, I. and Kendall, C.G.St.C. (1977). An interpretation of the depositional setting of some deep-water Jurassic carbonates of the Central High Atlas Mountains, Morocco. In: Cook, H.E. and Enos,

P. (eds.) *Deep water carbonate environments. Spec. Publ. Soc. econ. Paleontol. Mineral.* **25**, 249–261.

Evans, I., Kendall, C.G.St.C. and Butler, J.C. (1977). Genesis of Liassic shallow and deep water rhythms central High Atlas Mountains, Morocco. *J. sediment. Petrol.* **47** (1), 120–128.

Evans, P.R. (1981). The petroleum potential of Australia. *J. Petrol. Geol.* **4** (2), 123–146.

Exon, N.F., Rad, U. von and Stackelberg, U. von (1982). The geological development of the passive margins of the Exmouth Plateau off Northwest Australia. *Mar. Geol.* **47** (1–2), 131–152.

Exon, N.F. and Senior, B.R. (1976). The Cretaceous of the Eromanga and Surat Basins. *B.M.R. J. Aust. Geol. Geophys.* **1** (1), 33–50.

Exon, N.F. and Willcox, J.B. (1978). Geology and petroleum potential of Exmouth plateau area off western Australia. *Bull. Am. Assoc. Petrol. Geol.* **62** (1), 40–72.

Exon, N.F. and Willcox, J.B. (1980). The Exmouth Plateau: Stratigraphy, structure and petroleum potential. *Bull. Bur. Miner. Resour.* **199**, 52pp.

Exton, J. (1979). Mesozoic facies development and distribution: late Triassic-middle Jurassic. In: Wilson, R.C.L. and Exton, J. *Petrol. Explor. Soc. G.B. excursion to the Mesozoic of the Lusitanian Basin, west central Portugal, Sept. 9–15, 1979.* Dept. Earth Sci., the Open University, 50–58.

Eynon, G. (1981). Basin development and sedimentation in the Middle Jurassic of the northern North Sea. In: Illing, L.V. and Hobson, G.D. (eds.) *Petroleum geology of the continental shelf of north-west Europe. Proc. 2nd conference, London, March 1980, Inst. Petrol. Heydon and Son.* 196–204.

Fabre-Taxy, S. and Philip, J. (1966). Biostratigraphie et paléoécologie des peuplements à Huitres et des formations saumâtres du Santonien provencal. *Bull. Soc. géol. Fr. Sér. 7.* **8** (5), 702–711.

Fabricius, F.H. (1966). Beckensedimentation und Riffbildung an der Wende Trias/Jura in den Bayerischtiroler Kalkalpen. *Int. Sediment. Petrogr. Ser.* **9**. E.J. Brill. Leiden. 143pp + plates.

Faill, R.T. (1973). Tectonic development of the Triassic Newark-Gettysburg basin in Pennsylvania. *Bull. geol. Soc. Am.* **84** (3), 725–740.

Farquharson, G.W. (1983). The Nordenskjold Formation of the northern Antarctic Peninsula: an Upper Jurassic radiolarian mudstone and tuff sequence. *Bull. Br. Antarct. Surv.* **60**, 1–22.

Fassett, J.E. and Hinds, J.S. (1971). Geology and fuel resources of the Fruitland Formation and Kirtland Shale of the San Juan basin, New Mexico and Colorado. *Prof. pap. U.S. geol. Surv.* **676**, 76pp.

Faucher, B. and Savoyat, E. (1973). Esquisse géologique des Andes de L'Equateur. *Rev. Géogr. phys. Géol. dyn.* **15** (1–2), 115–142.

Faugeres, J-C. (1981). Evolution structurale d'un bassin atlantico-mésogéen de la marge africaine: les rides sud-rifaines (Maroc). *Bull. Soc. géol. Fr. Sér. 7.* **23** (3), 229–244.

Faugeres, J-C., Lang, J., Lucas, G. and Perriaux, J. (1970). Contribution à l'étude des facies détritiques du Crétacé Moyen du Vercors. *Géol. Alpine.* **46**, 77–86.

Faupl, P. (1979). Turbiditserien in den Kreideablagerungen des Ostalpins und ihre paläogeographische Bedeutung. In: Wiedmann, J. (ed.) *Aspekte der Kreide Europas, Proc. 1st Symp. German Cretaceous, Münster, April 1978, Int. Union geol. Sci., Ser. A.* **6**, E. Schweizerbart'sche Verlagsbuchhandlung, Stuttgart, 403–412.

Fay, M. and Groschke, M. (1982). Die Mittel Jura-Sandsteine in

Niederbayern-Lithologie, Stratigraphie, Paläogeographie. *Neues Jahrb. Geol. Paläontol. Abhandlungen.* **163** (1), 23–48.

Fazzuoli, M. (1980a). Frammentazione ed annegamento della piattforma carbonatica del Calcare massiccio (Lias Inferiore) nell'area Toscana. In: *Atti del 70th Congr. Soc. geol. Ital., L'evoluzione tettonico-sedimentaria dell'Appennino settentrionale, Siena, Oct. 1980. Mem. Soc. geol. Ital.* **21**, 181–191.

Fazzuoli, M. (1980b). Considerazioni preliminari sul calcare selcifero della val di Lima (Giurassico Superiore) Toscana nord-occidentale. In: *Atti del 70th Congr. Soc. geol. Ital., L'evoluzione tettonico-sedimentaria dell'Appennino settentrionale, Siena, Oct. 1980. Mem. Soc. geol. Ital.* **21**, 193–201.

Fazzuoli, M. and Pirini Radrizzani, C. (1981). Lithofacies characteristics of the 'Rosso Ammonitico' limestone in the south-western Tuscany (Italy). In: Farinacci, A. and Elmi, S. (eds.) *Proc. Rosso Ammonitico Symp., Rome, June 1980.* Edizioni Tecnoscienza, Rome, 409–417.

Feininger, T. and Bristow, C.R. (1980). Cretaceous and Paleogene geologic history of coastal Ecuador. *Geol. Rdsch.* **69** (3), 849–874.

Felber, J., Hoffmann, R.,Leinfedler, R., Oschmann, W. and Ziegler, R. (1982). Biofaziesanalyse randmariner Ablagerungsbereiche im Oberen Jura Portugals, II: Paläogeographie und Faziesverteilung in der Serra da Arrabida. *Neues Jahrb. Geol. Paläontol. Abhandlungen.* **163** (3), 301–330.

Ferriere, J. and Mariotti, A. (1971). Précisions sur la paléogéographie des Alpes méridionales: la 'sous-zone subtridentine' (Val di Non, Ouest des Dolomites de Brenta, provinces de Trento et Bolzano, Italie). *Bull. Soc. géol. Fr. Sér. 7.* **13** (3–4), 298–314.

Ferry, S. (1978). Les 'Calcaires à débris' Barremo-Aptiens de la Drôme Vocontienne (France sud-est): des cônes d'épandage bioclastique en eau profonde. In: *Les sédiments: leur histoire, leur environnement et leur devenir, Livre Jubilaire Jacques Flandrin. Doc. Lab. Géol. Fac. Sci. Lyon. Hors Sér.* **4**, 273–304.

Ferry, S. (1979). Les turbidites bioclastiques péri-urgoniennes: modalités de transport du sable calcaire depuis les plates-formes jusque dans le bassin profond. Exemple de la cuvette vocontienne (S.E. de la France). In: *L'Urgonien des Pays Méditerranéens. Géobios. Spéc. Mém.* **3**, 149–158.

Feuillee, P. (1967). Le Cénomanien des Pyrénées basques aux Asturies. *Mém. Soc. géol. Fr.* **46**, 343pp.

Feuillee, P. (1971). Les calcaires biogéniques de l'Albien et du Cénomanien Pyrénéo-Cantabrique: problèmes d'environnement sédimentaire. *Palaeogeogr. Palaeoclimatol. Palaeoecol.* **9** (4), 277–311.

Feuillee, P. and Rat, P. (1971). Structures et paléogéographies Pyrénéo-Cantabriques. In: *Proc. Symp. Histoire structurale du Golfe de Gascogne, Rueil-Malmaison, Déc. 1970, Publ. Inst. Fr. Pét., collection colloq. et séminaires.* **22** (2), Editions Technip, Paris, V-1, 1–48.

Feuillee, P., Villanova, M. and Winnock, E. (1973). La dynamique des fosses à turbidites et de leur contenu sédimentaire dans le système pyrénéen. *Bull. Soc. géol. Fr. Sér. 7.* **15** (1), 61–76.

Fischer, W. (1964). Die Jurensis-Schichten (Oberes Toarcien) im Vorland der Schwäbischen Alb. *Neues Jahrb. Geol. Paläontol. Abhandlungen.* **120**, 81–106.

Fisher, W.L. and Rodda, P.U. (1969). Edwards Formation (Lower Cretaceous) Texas: dolomitization in a carbonate platform system. *Bull. Am. Assoc. Petrol. Geol.* **53** (1), 55–72.

Fleming, C.A. (1962). New Zealand biogeography: a palaeontologist's approach. *Tuatara.* **10**, 53–108.

Fleury, J-J. (1968). La marge orientale du cause Mégéan et du cause Noir. Stratigraphie et paléogéographie du Lias et de l'Aalenien-Bajocien dans la région de Meyrueis (Lozère). *Bull. Soc. géol. Fr. Sér. 7.* **10** (6), 645–652.

Flexer, A. (1968). Stratigraphy and facies development of Mount Scopus Group (Senonian-Paleocene) in Israel and adjacent countries. *Israel J. earth. Sci.* **17**, 85–114.

Flexer, A. (1971). Late Cretaceous palaeogeography of northern Israel and its significance for the Levant geology. *Palaeogeogr. Palaeoclimatol. Palaeoecol.* **10** (4), 293–310.

Floquet, M. (1978). La sédimentation de plate-forme au Crétacé Supérieur dans la Vieille Castille (Espagne): évolution verticale, variation horizontale; implications paléogéographiques. *Bull. Soc. géol. Fr. Sér. 7.* **20** (5), 779–783.

Floquet, M. and Rat, P. (1975). Un exemple d'interrelation entre socle, paléogéographie et structure dans l'arc Pyrénéen Basque: La Sierra d'Aralar. *Rev. Géogr. phys. Géol. dyn.* **17** (5), 497–512.

Fois, E. and Gaetani, M. (1980). The northern margin of the Civetta Buildup. Evolution during the Ladinian and the Carnian. *Riv. Ital. Paleontol. Stratigr.* **86** (3), 469–542.

Fontaine, J-M. (1981). *La plate-forme arabe et sa marge passive au Mésozoïque: l'exemple d'Hazro (S.E. Turquie).* Unpubl. PhD. Thesis, Univ. Paris-Sud. Centre D'Orsay. 270pp.

Foose, R.M. and Manheim, F. (1975). Geology of Bulgaria: a review. *Bull. Am. Assoc. Petrol. Geol.* **59** (2), 303–335.

Ford, A. and Houbolt, J.J.H.C. (1963). The microfacies of the Cretaceous of western Venezuela. *Int. sediment. petrogr. Ser.* **6**, E.J. Brill, Leiden, 55pp + plates.

Forgotson, J.M., Jnr. (1954). Regional stratigraphic analysis of Cotton Valley group of Upper Gulf Coastal Plain. *Bull. Am. Assoc. Petrol. Geol.* **38** (12), 2476–2499.

Forgotson, J.M. (1963). Depositional history and paleotectonic framework of Comanchean Cretaceous Trinity stage, Gulf coast area. *Bull. Am. Assoc. Petrol. Geol.* **47** (1), 69–103.

Forster, H., Hagedorn, H. and Haars, W. (1975). Unterkretazische terrestrische Sedimente und ihre Diagenese im Shirkuh-Massiv (Zentral Iran). *Neues Jahrb. Geol. Paläontol. Monatshefte.* **1975** (5), 279–292.

Forster, R. (1975). Die geologische Entwicklung von Süd-Mozambique seit der Unterkreide und die Ammoniten-Fauna von Unterkreide und Cenoman. *Geol. Jahrb.* B12, 3–324.

Forster, R. (1975). The geological history of the sedimentary basin of southern Mozambique, and some aspects of the origin of the Mozambique Channel. *Palaeogeogr. Palaeoclimatol. Palaeoecol.* **17** (4), 267–287.

Forster, R. 1981. Mozambique. In: Reyment, R.A. and Bengtson, P. (eds.) *Aspects of Mid-Cretaceous Regional Geology.* Academic Press, London, 161–174.

Forster, R. and Scholz, G. (1979). *Salaziceras nigerianum* n. sp. from southeast Nigeria: Faunal evidence for an open seaway between the northern and southern Atlantic in Late Albian times. *Neues Jahrb. Geol. Palaeontol. Monatshefte.* **1979** (2), 109–119.

Fouch, T.D., Lawton, T.F., Nichols, D.J., Cashion, W.B. and Cobban, W.A. (1983). Patterns and timing of synorogenic sedimentation in Upper Cretaceous rocks of central and northeast Utah. In: Reynolds, M.W. and Dolly, E.D. (eds.) *Mesozoic Paleogeography of the*

Bibliography

West-Central United States, Proc. 2nd Rocky Mt. Paleogeogr. Symp., Denver, Colorado. Rocky Mt. Sect., Soc. econ. Paleontol. Mineral. 305–336.

Fourcade, E., Azema, J., Chabrier, G., Chauve, P., Foucault, A. and Rangheard, Y. (1977). Liaisons paléogéographiques au Mésozoïque entre les zones externes bétiques, Baléares, Corso-Sardes et Alpines. *Rev. Géogr. phys. Géol. dyn.* **19** (4), 377–388.

Francis, D.R. (1957). Jurassic stratigraphy of Williston Basin area. *Bull. Am. Assoc. Petrol. Geol.* **41** (3), 367–398.

Franks, S. and Nairn, A.E.M. (1973). The equatorial marginal basins of west Africa. In: Nairn, A.E.M. and Stehli, F.G. (eds.) *The ocean basins and margins*, **1**, *The South Atlantic.* Plenum Press, London, 301–350.

Freeman, T. (1972). Sedimentology and dolomitization of Muschelkalk carbonates (Triassic), Iberian range, Spain. *Bull. Am. Assoc. Petrol. Geol.* **56** (3), 434–453.

Freimoser, M. (1972). Zur Stratigraphie, Sedimentpetrographie und Faziesentwicklung der südostbayerischen Flyschzone und des Ultrahelvetikums zwischen Bergen/Obb. und Salzburg. *Geologica Bav.* **66** (1), 7–91.

Freund, R. (1965). Upper Cretaceous reefs in northern Israel. *Israel J. earth Sci.* **14**, 108–121.

Freund, R. and Raab, M. (1969). Lower Turonian ammonites from Israel. *Spec. Pap. Palaeontol.* **4**, 83pp.

Frey, M. (1968). Quartenschiefer, Equisetenschiefer und germanischer Keuper – ein lithostratigraphischer Vergleich. *Eclog. geol. Helv.* **61** (1), 141–156.

Freyberg, B. (1959). Zur Stratigraphie und Paläogeographie des Doggersandsteins in der Fränkischen Alb. *Erlanger geol. Abhandlungen.* **30**, 68pp.

Freyberg, B. (1963). Der Obere Malm an der unteren Altmuhl: Nebst einer Studie über das Riff-Problem. *Erlanger geol. Abhandlungen.* **49**, 38pp.

Freyberg, B. (1966). Die Faziesverband im Unteren Malm Frankens: Ergebnisse der Stromatometrie. *Erlanger geol. Abhandlungen.* **62**, 122pp.

Freyberg, B. (1968). Übersicht über den Malm der Altmühl-Alb. *Erlanger geol. Abhandlungen.* **70**, 40pp.

Freytet, P. (1971a). Les dépôts continentaux et marins du Crétacé Supérieur et des couches de passage à l'Eocène en Languedoc. *Bull. Bur. Rech. géol. minières. sér. 2, Sect. I,* **4**, 1–54.

Freytet, P. (1971b). Le Languedoc au Crétacé Supérieur et à l'Eocène Inférieur: évolution des principaux éléments structuraux (rides et sillons), migration des aires de sédimentation, rôle des phases précoces dans la tectogénèse. *Bull. Soc. géol. Fr. Sér. 7.* **13** (5–6), 464–474.

Freytet, P. (1973). Edifices récifaux développés dans un environnement détritique: exemple des biostromes à *Hippurites* (rudistes) du Senonien Inférieur du sillon languedocien (région de Narbonne, Sud de la France). *Palaeogeogr. Palaeoclimatol. Palaeoecol.* **13** (1), 65–76.

Frisch, J. (1975). Sedimentologische, lithofazielle und paläogeographische Untersuchungen in den Reichenhaller Schichten und im Alpinen Muschelkalk der Nördlichen Kalkalpen zwischen Lech und Isar. *Jahrb. geol. Bundesanst.* **118**, 75–117.

Fruth, I. and Scherreiks, R. (1982). Hauptdolomit (Norian) – stratigraphy, palaeogeography and diagenesis. *Sediment. Geol.* **32** (3), 195–231.

Fuchs, W. and Stradner, H. (1967). Die Foraminiferenfauna und Nannoflora eines Bohrkernes aus dem höheren Mittel-Alb der Tiefbohrung Delft 2, (NAM), Niederlande. *Jahrb. geol. Bundesanst.* **110** (2), 245–341.

Fujita, K. (1978). Pre-Cenozoic tectonic evolution of north-east Siberia. *J. Geol.* **86** (2), 159–172.

Fulop, J. (1971). Les formations Jurassiques de la Hongrie. In: Vegh-Neubrandt, E. (ed.) *Colloque du Jurassique Méditerranéen, Budapest, 1969, Ann. Inst. geol. publici Hung.* 54 (2), 31–46.

Fursich, F.T. and Sykes, R.M. (1977). Palaeobiogeography of the European Boreal realm during Oxfordian (Upper Jurassic) times: a quantitative approach. *Neues Jahrb. Geol. Palaontol. Abhandlungen.* **155**, 137–161.

Furst, M. (1965). Die Oberkreide-Paleozan Transgression im östlichen Fezzan. *Geol. Rdsch.* **54** (2), 1060–1088.

Gaetani, M., Fois, E., Jadoul, F. and Nicora, A. (1981). Nature and evolution of Middle Triassic carbonate buildups in the Dolomites (Italy). In: Cita, M.B. and Ryan, W.B.F. (eds.) Carbonate platforms of the passive type continental margins: present and past. *Mar. Geol.* **44** (1–2), 25–58.

Gajewska, I. (1978). Zarys rozwoju prac badawczych i pogladow na stratygrafie kajpru wpolnocno - zachodniej Polsce. In: Modzelewska, B (ed.) *Stratygrafia kajpru w Polsce. Pr. Inst. Geol. Warzawa.* **87**, 5–59.

Gall, H., Huttner, R. and Muller, D. (1977). Erläuterungen zur geologischen Karte des Reises 1:50,000. *Geologica Bav.* **76**, 191pp.

Gall, H., Muller, D. and Yamani, A. (1973). Zur Stratigraphie und Paläogeographie der Cenoman-Ablagerungen auf der südwestlichen Frankenalb (Bayern). *Neues Jahrb. Geol. Paläontol. Abhandlungen.* **143** (1), 1–22.

Gallois, R.W. and Cox, B.M. (1977). The stratigraphy of the Middle and Upper Oxfordian sediments or Fenland. *Proc. Geol. Assoc. Lond.* **88** (4), 207–228.

Ganev, M. (1974). Stand der Kenntnisse über die Stratigraphie der Trias Bulgariens. In: Zapfe, H. (ed.) *The Stratigraphy of the Alpine-Mediterranean Triassic, Proc. Symp. Wien, May 1973,* Springer-Verlag, Wien, 93–96.

Garcia-Hernandez, M. (1979). Les facies urgoniens pendant la sédimentation Barremo-Albienne dans les Sierras de Cazorla et du Segura (zone prébétique. Espagne). In: *L'Urgonien des Pays Méditerranéens. Géobios. Spéc. Mém.* **3**, 57–70.

Garrido-Megías, A. (1975). El Cretácico Superior de la zona central de la vertiente surpirenaica, dentro del contexto de la Cadena Pirenaica. In: *Primer coloquio de estratigrafía y paleogeografía del Cretácico de España. Bellaterra-Tremp, November 1973, Publ. Empresa Nac. Adaro. Invest. Mineras, Ser. 7, Trabajos de congresos y reuniones., Madrid,* 1, 103–116.

Garrido-Megías, A. and Morales, J.V. (1977). El Trías germánico en España: paleogeografía y estudio secuencial. Proc. Coloquio de estratigrafía y paleogeografía del Triásico y Pérmico de España, Madrid, 1977 *Cuad. Geol. Ibérica.* **4**, 37–56.

Gatrall, M., Jenkyns, H.C. and Parsons, C.F. (1972). Limonitic concretions from the European Jurassic, with particular reference to the 'snuff-boxes' of southern England. *Sedimentol.* **18**, 79–103.

Gautier, D.L. (1981). Lithology, reservoir properties and burial history of portion of Gammon Shale (Cretaceous), southwestern North Dakota. *Bull. Am. Assoc. Petrol. Geol.* **65** (6), 1146–1159.

Gealey, W.K. (1977). Ophiolite obduction and geologic evolution of the Oman Mountains and adjacent areas. *Bull. geol. Soc. Am.* **88** (8), 1183–1191.

Gelati, R., Cascone, A. and Paggi, L. (1982). Le unita' stratigrafiche Aptiano-Maastrichtiane delle Prealpi bergamasche. *Riv. Ital. Palaeontol. Stratigr.* **88** (1), 73–112.

Gelmini, R. and Mantovani, M.P. (1982). La successione Triassica di Col de Serna (Corsica Settentrionale). *Riv. Ital. Palaeontol. Stratigr.* **88** (1), 11–20.

George, T.N. (1970). *South Wales. British Regional Geology. 3rd Edn.* H.M.S.O. London.

Geyer, O.F. (1973). Das präkretazische Mesozoikum von Kolumbien. *Geol. Jahrb.* **B5**, 156pp.

Geyer, O.F. (1974). Der Unterjura (Santiago-Formation) von Ekuador. *Neues Jahrb. Geol. Paläontol. Monatshefte* **1974** (9), 525–541.

Geyer, O.F. (1979). Zur Paläogeographie mesozoischer Ingressionen und Transgressionen in Kolumbien. *Neues Jahrb. Geol. Paläontol. Monatshefte.* **1979** (6), 349–368.

Gibergy, P. and Sarrot-Reynauld, J. (1965). Stratigraphie et variations de facies du Crétacé Inférieur aux environs de Saillans (Drôme). In: *Proc. Colloque sur le Crétacé Inférieur, Lyon, Sept. 1963, Mém. Bur. Rech. géol. minières.* **34**, 753–768.

Gill, J.R. and Cobban, W.A. (1965). Stratigraphy of the Pierre Shale, Valley City and Pembina Mountain areas, North Dakota. *Prof. pap. U.S. geol. Surv.* **392-A**, 20pp.

Gill, J.R. and Cobban, W.A. (1973). Stratigraphy and geologic history of the Montana Group and equivalent rocks, Montana, Wyoming and North and South Dakota. *Prof. pap. U.S. geol. Surv.* **776**, 37pp.

Gill, J.R., Cobban, W.A. and Schultz, L.G. (1972). Stratigraphy and composition of the Sharon Springs Member of the Pierre Shale in western Kansas. *Prof. pap. U.S. geol. Surv.* **1728**, 50pp.

Gillet, S. (1965). Les Trigonies du Crétacé Inférieur. In: *Proc. Colloque sur le Crétacé Inférieur, Lyon, Sept. 1963, Mém. Bur. Rech. géol. minières.* **34**, 399–408.

Girard, D., Beck, C., Stephan, J-F., Blanchet, R. and Maury, R. (1982). Pétrologie, géochimie et signification géodynamique de quelques formations volcaniques Crétacées péri-caraïbes. *Bull. Soc. géol. Fr. Sér. 7.* **24** (3), 535–544.

Given, M.M. (1977). Mesozoic and Early Cenozoic geology of offshore Nova Scotia. *Bull. Can. Pet. Geol.* **25** (1), 63–91.

Glaister, R.P. (1959). Lower Cretaceous of southern Alberta and adjoining areas. *Bull. Am. Assoc. Petrol. Geol.* **43** (3), 590–640.

Glashoff, J. (1964). Ostrakoden-Faunen und Paläogeographie im Oxford N.W. Europas. *Palaeontol. Z. Berlin.* **38 (1–2)**, 28–65.

Gnaccolini, M. and Martinis, B. (1974). Nuove ricerche sulle Formazioni Calcaree Giurassico-Cretaciche della regione compresa tra le valli del Natisone e del Piave. In: *Contributi stratigrafici e paleogeografici sul Mesozoico della Tetide. Mem. Riv. Ital. Paleontol. Stratigr.* **14**, 5–109.

Gocev, P.M. (1976). L'évolution géotectonique du megabloc Bulgare pendant le Trias et le Jurassique. *Bull. Soc. géol. Fr. Sér. 7.* **18** (2), 209–216.

Goldberg, M. and Friedman, G.M. (1974). Paleoenvironments and paleogeographic evolution of the Jurassic system in southern Israel. *Bull. geol. Surv. Israel.* **61**, 44pp.

Gómez, J.J. (1979). El Jurasico en facies carbonatadas del sector levantino de la Cordillera Ibérica. *Seminarios de estratigrafia. Ser.*

monografia. **4**, 683pp.

Gordon, W.A. (1973). Marine life and ocean surface currents in the Cretaceous. *J. Geol.* **81** (3), 269–284.

Gordon, W.A. (1974). Physical controls on marine biotic distribution in the Jurassic period. In: Ross, A. (ed.) Paleogeographic provinces and provinciality, *Spec. Publ. Soc. econ. Paleontol. Mineral.* **21**, 136–147.

Gorter, J.D. (1978). Triassic environments in the Canning Basin, Western Australia. *B.M.R. J. Aust. Geol. Geophys.* **3** (1), 25–33.

Gracianksy, P.C.de., Brosse, E., Deroo, G., Herbin, J-P., Mondaert, L., Muller, C., Sigal, J. and Schaaf, A. (1982). Les formations d'âge de l'Atlantique nord et leur matière organique: Paléogéographie et milieux de dépôt. *Rev. Inst. Fr. Pét.* **37** (3), 275–337.

Graciansky, P.C.de., and Lemoine, M. (1980). Paléomarge de la Téthys dans les Alpes occidentales du massif Central francais aux ophiolites liguro-piémontaises. *Géol. Alpine.* **56**, 119–147.

Graindor, M.J. (1963). Le socle armoricain et les contre-coups alpins. In: Durand-Delga, M. (ed.) *Livre à la Mémoire du Professeur Paul Fallot: L'évolution paléogéographique et structurale des domaines méditerranéens et Alpins d'Europe. Mém. Soc. géol. Fr. Hors Sér.l,* **2**, 187–202.

Grandjacquet, C. and Mascle, G. (1973). The structure of the Ionian Sea, Sicily and Calabria-Lucania. In: Nairn, A.E.M., Kanes, W.H. and Stehli, F.G. (eds.) *Ocean basins and margins,* **4B**, *The west Mediterranean,* Plenum Press, New York, 257–330.

Grasso, M. and Lentini, F. (1982). Sedimentary and tectonic evolution of the eastern Hyblean Plateau (southeastern Sicily) during late Cretaceous to Quaternary Time. *Palaeogeogr. Palaeoclimatol. Palaeoecol.* **39** (3–4), 261–280.

Graupner, A. von (1968). Gliederung und Lagerstätten des kohleführenden Wealdens im Hilsgebiet. *Geol. Jahrb.* **85**, 165–192.

Gravesen, P., Rolle, F. and Surlyk, F. (1982). Lithostratigraphy and sedimentary evolution of the Triassic, Jurassic and Lower Cretaceous of Bornholm, Denmark. *Dan. geol. unders., Ser. B.* **7**, 51pp.

Green, A.R. and Kaplan, A. (1982). The evolution of sedimentary basins in the circum-Arctic. In: Proc. Offshore Northern Seas Conf., Stavanger, Norway, Aug. 1982, E, The geological framework and hydrocarbon potential of basins in northern seas, **E/1**, 1–53.

Grigyalis, A.A., Martinene, R.A., Akimez, W.S., Gluschko, W.W., Wischnjakow, I.B., Poluchtowitsch, B.M., Sonjkowski, Ju.N., Gawrmischin, U.I. and Pasternak, S.I. (1978). Paläogeographie des Albs auf dem westlichen Territorium des europäischen Teils der UdSSR. *Z. angew. Geol. DDR.* **24** (11), 451–456.

Grigyalis, A.A. and Rotkite, L.M. (1972). Kimmeridgian stage of the Baltic region. *Dokl. Acad. Sci. USSR.* **198**, 110–112.

Groiss, J.T. (1970). Feinstratigraphische, ökologische und zoogeographische Untersuchungen der Foraminiferen-Faunen im Oxford der Franken-Alb. *Erlanger geol. Abhandlungen.* **81**, 83pp.

Grubic, A. (1980). Yougoslavie. In: *Géologie des pays européens: Espagne, Grèce, Italie, Portugal, Yougoslavie. Com. Nat. Fr. géol., 26th Int. geol. congr., Paris,* **1980.** 287–342.

Grupo Espanol de Trabajo Proyecto No. 58 Mid-Cretaceous Events (I.G.C.P.) (1982). *El Cretacico de Espana.* Universidad Complutense de Madrid, 680pp.

Gruss, H. and Theinhaus, R. (1969). Paläogeographie und Enstehung der Eisenerze der Ober-Aalenium (Dogger B) Nordwestdeutschlands. In: Bottke, H., Dengler, H., Finkenwirth, A. *et al., Sammelwerk*

Deutsche Eisenerzlagerstätten: II Eisenerze im Deckegebirge (Post-varistikum), 1, Die Marin-sedimentaren Eisenerze des Jura in Nordwest-deutschland. Beih. geol. Jahrb. **79**, 167–172.

Gudden, J. (1972). Die Bildung und Erhaltung der Oberpfälzer Kreide-Eisenerzlagerstätten in Abhängigkeit von Biegungs – und Bruchtektonik. *Geologica Bav.* **65**, 107–125.

Guendon, J.L., Parron, C. and Triat, J.M. (1982). Incidences des altérations Crétacées sur la notion de Sidérolithique dans le sud-est de la France. *Bull. Soc. géol. Fr. Sér. 7.* **25** (1), 41–50.

Guiraud, R. (1975). L'évolution post-Triasique de l'avant-pays de la chaîne Alpine en Algérie d'après l'étude du bassin du Hodna et des régions Voisines. *Rev. Géogr. phys. Géol. dyn.* **17** (4), 427–446.

Gutnic, M., Monod, 0., Poisson, A. and Dumont, F-F. (1979). Géologie des Taurides occidentales (Turquie). *Mém. Soc. géol. Fr. Hors Sér.* **137**, 109pp.

Gwinner, M.P. (1971). *Geologie der Alpen: Stratigraphie, Paläeogeographie, Tektonik.* E. Schweizerbart'sche Verlagsbuchhandlung, Stuttgart, 477pp.

Gwinner, M.P. (1976). Origin of Upper Jurassic limestones from the Swabian Alb (south west Germany). *Contrib. Sedimentol. Stuttgart.* **5**, 75pp.

Haanstra, U. (1963). A review of mesozoic geological history in the Netherlands. *Verh. K. Ned. geol. mijnb. Genoot. Geol. Ser.* **21**, 35–55.

Haas, J. (1979). A felsökréta ugodi mészkö formáció a Bakonyban. *Ann. Inst. geol. publici Hung.* **61**, 171pp.

Hahn, L. (1976). The stratigraphy and paleogeography of the non-marine Mesozoic deposits in northern Thailand. *Geol. Jahrb.* **B21**, 155–169.

Hahn, L. (1982). The Triassic in Thailand. In: *Proc. Symp. Triassic global geology, Wurzburg, Feb. 1982, Geol. Rdsch.* 71 (3), 1041–1056.

Hakenberg, M. (1978). Paleotektonika i paleogeografi ponocnef części niecki Miechowskiej w Albie i Cenomanie. *Stud. geol. Pol.* 58, 104pp.

Halbouty, M.T. (1966). Stratigraphic-trap possibilities in Upper Jurassic rocks, San Marcos Arch, Texas. *Bull. Am. Assoc. Petrol. Geol.* **50** (1), 3–24.

Hallam, A. (1965). Observations on marine Lower Jurassic stratigraphy of north America, with special reference to United States. *Bull. Am. Assoc. Petrol. Geol.* **49** (9), 1485–1501.

Hallam, A. (1967). An environmental study of the Upper Domerian and Lower Toarcian in Great Britain. *Philos. Trans. R. Soc. Lond., Ser. B.* **252**, 393–445.

Hallam, A. (1969). Faunal realms and facies in the Jurassic. *Palaeontol.* **12** (1), 1–18.

Hallam, A. (1971a). Jurassic faunal provinces in relation to sedimentary facies and palaeogeography. In: Middlemiss, F.A., Rawson, P.F., and Newall, (eds.) *Faunal provinces in space and time. Spec. Issue. Geol. J.* **4**, 129–152.

Hallam, A. (1971b). Facies analysis of the Lias in West Central Portugal. *Neues Jahrb. Geol. Palaontol. Abhandlungen.* **139** (2), 226–265.

Hallam, A. (1971c). Mesozoic geology and the opening of the North Atlantic. *J. Geol.* **79** (2), 129–157.

Hallam, A. (1975). *Jurassic environments.* Cambridge Earth Sci. Scr., Cambridge Univ. Press, 269pp.

Hallam, A. (1981). The end – Triassic bivalve extinction event. *Palaeogeogr. Palaeoclimatol. Palaeoecol.* **35** (1), 1–44.

Hallam, A. and El Shaarawy, Z. (1982). Salinity reduction of the end – Triassic sea from the Alpine region into northwestern Europe.

Lethaia. **15** (2), 169–178.

Hallam, A. and Sellwood, B.W. (1975). Middle Mesozoic sedimentation in relation to tectonics in the British area. In: Finstad, K.G. and Selley, R.C. (eds.) *Jurassic northern North Sea Symposium, Stavanger, Sept. 1975, Nor. Pet. Foren., Oslo, JNNSS/4.* 32pp. (Also published in *J. Geol.* **84**, 301–32, (1976).)

Hamblin, A.P. and Walker, R.G. (1979). Storm-dominated shallow marine deposits: the Fernie-Kootenay (Jurassic) transition, southern Rocky Mountains. *Can. J. Earth Sci.* **16** (9), 1673–1690.

Hamilton, W. (1969). Mesozoic California and the underflow of Pacific mantle. *Bull. geol. Soc. Am.* **80** (12), 2409–2430.

Hammuda, O.S. (1980). Sediments and palaeogeography of the Lower Campanian sand bodies along the southern tip of Ad Daffah-Al Wahah ridge, Sirte basin. In: Salem, M.J. and Busrewil, M.T. (eds.) *The geology of Libya,* **2**. *Proc. 2nd Symp. Tripoli,* **1978**. Academic Press, London, 509–520.

Hancock, J.M. (1975). The sequence of facies in the Upper Cretaceous of northern Europe compared with that in the Western Interior. In: Caldwell, W.G.E. (ed.) *The Cretaceous System in the Western Interior of North America. Proc. Int. Symp. Univ. Saskatchewan, Saskatoon, May 1973, Spec. Pap. geol. Assoc. Can.* **13**, 83–118.

Hancock, J.M. (1975). The petrology of the Chalk. *Proc. Geol. Assoc. Lond.* **86** (4), 499–535.

Hantzpergue, P. and Maire, P. (1981). Les plates-formes de l'Aunis et de l'Angoumois au Jurassique Supérieur: caractères sédimentologiques et paléogéographie. *Bull. Soc. géol. Fr. Sér. 7.* **23** (5), 493–500.

Harrington, H.J. (1962). Palaeogeographic development of South America. *Bull. Am. Assoc. Petrol. Geol.* **46** (10), 1773–1814.

Harrison, J. (1969). A review of the sedimentary history of the Island of New Guinea. *Aust. Petrol. Explor. Assoc. J.* **9** (2), 41–48.

Harsch, W. (1970). Zur Sedimentologie und Paläogeographie der Raibler Schichten (Ober-Trias) im Westteil der Nördlichen Kalkalpen. *Z. Deutsch. geol. Ges.* **121**, 253–272.

Hart, B.B.'t. von (1969). Die Oberjura – und Unterkreide – Sedimentation in den nördlichen und östlichen Niederlanden. *Erdöl und Kohle. Erdgas. Petrochem.* **22** (5), 253–261.

Hart, M.B. (1976). The mid-Cretaceous succession of Orphan Knoll (northwest Atlantic): micropalaeontology and palaeo-oceanographic implications. *Can. J. Earth Sci.* **13** (10), 1411–1421.

Hart, M.B. and Bigg, P.J. (1981). Anoxic events in the late Cretaceous chalk seas of north-west Europe. In: Neale, J.W. and Brasier, M.D. (eds.) *Microfossils from recent and fossil shelf seas, Proc. Mtg. British Micropalaeont. Soc. July 1980, British Micropaleontol. Soc. Ser.* Ellis Horwood, Chichester, 177–185.

Hart, M.B. and Tarling, D.H. (1974). Cenomanian palaeogeography of the North Atlantic and possible Mid-Cenomanian eustatic movements and their implications. *Palaeogeogr. Palaeoclimatol. Palaeoecol.* **15** (2), 95–108.

Hattin, D.E. (1975). Stratigraphy and depositional environment of Greenhorn Limestone (Upper Cretaceous) of Kansas. *Bull. Kansas geol. Surv.* **209**, 128pp.

Hattin, D.E. (1982). Stratigraphy and depositional environment of Smoky Hill Chalk Member, Niobrara Chalk (Upper Cretaceous) of the type area, western Kansas. *Bull. Kansas geol. Surv.* **225**, 108pp.

Haun, J.D. and Kent, H.C. (1965). Geologic history of Rocky Mountain region. *Bull. Am. Assoc. Petrol. Geol.* **49** (11), 1781–1800.

Hay, W.W., Behensky, Jnr, J.F., Barron, E.J. and Sloan II, J.L. (1982). Late Triassic-Liassic paleoclimatology of the proto-central North Atlantic rift system. *Palaeogeogr. Palaeoclimatol. Palaeoecol.* **40** (1–3), 13–3.

Hayes, P.T. (1970). Cretaceous paleogeography of southeastern Arizona and adjacent areas. *Prof. pap. U.S. geol. Surv.* **658-B**, 42pp.

Heaton, R.L. (1950). Late Paleozoic and Mesozoic history of Colorado and adjacent areas. *Bull. Am. Assoc. Petrol. Geol.* **34** (8), 1659–1698.

Helal, A.H. (1965). Jurassic spores and pollen grains from the Kharga Oasis, Western Desert, Egypt. *Neues Jahrb. Geol. Palaontol. Abhandlungen.* **123** (2), 160–166.

Heling, D. and Wurster, P. (1971). Deltaic sediments of the Keuper Basin in S.W. Germany. In: Muller, G. (ed.) *Sedimentology of parts of Central Europe, Guidebook to excursions. 8th Int. Sedimentol. Congr., Heidelburg, 1971.* Verlag Waldemar Kramer, Frankfurt, 133–146.

Henderson, G., Schiener, E.J., Risum, J.B., Croxton, C.A. and Andersen, B.B. (1981). The West Greenland Basin. In: Kerr, J.W. and Fergusson, A.J. (eds.) *Geology of the North Atlantic borderlands, Proc. Symp. St. John's, Newfoundland, June 1979, Mem. Can. Soc. Petrol. Geol.* **7**, 399–428.

Henry, J., Lanusse, R. and Villanova, M. (1971). Evolution du domaine marin Pyrénéen du Sénonien Supérieur à l'Eocène Inférieur. In: *Proc. Symp. Histoire structural du Golfe de Gascogne, Rueil-Malmaison, Déc. 1970, Publ. Inst. Fr. Pét., collection colloq. et séminaires.* **22** (1), Editions Technip, Paris, IV-7, 1–8.

Herak, M. (1974). Palaeogeographie und Fazies-Entwicklung wahrend der Trias in der Dinariden Kroatiens. In: Zapfe, H. (ed.) *The Stratigraphy of the Alpine-Mediterranean Triassic, Proc. Symp. Wien, May 1973.* Springer-Verlag, Wien, 101–106.

Herbert, C. (1980). Depositional development of the Sydney Basin. In: Herbert, C. and Helby, R. (eds.) *A guide to the Sydney Basin, Bull. Geol. Surv. New South Wales.* **26**, 10–53.

Hercogova, J. and Valecka, J. (1977). Coniac-Ablagerungen in der Kreide Ostböhmens. *Vesn. ustred. Ustavu geol. CAV.* **52**, 371–375.

Herm, D. (1979). Die süddeutsche Kreide – Ein Überblick. In: Wiedmann, J. (ed.) *Aspekte der Kreide Europas, Proc. 1st Symp. German Cretaceous, Münster, April 1978, Int. Union geol. Sci., Ser. A.* **6**, E. Schweizerbart'sche Verlagsbuchhandlung, Stuttgart, 85–106.

Hertle, A. (1962). Stratigraphie und Tektonik der Fränkischen Alb um Wissing. *Erlanger geol. Abhandlungen.* **45**, 41pp.

Herz, N. and Savu, H. (1974). Plate tectonic history of Romania. *Bull. geol. Soc. Am.* **85** (8), 1429–1440.

Hesemann, J. (1975). Geologie Nordrhein-Westfalens. *Bochumer Geogr. Arb., Sonderreihe.* 2, F. Schoningh, Paderborn, 416pp.

Hesse, R. and Butt, A. (1976). Paleobathymetry of Cretaceous turbidite basins of the east Alps relative to the calcite compensation level. *J. Geol.* **84** (5), 505–533.

Hillebrandt, A. (1981). Kontinentalverschiebung und die paläozoogeographischen Beziehungen des Südamerikanischen Lias. *Geol. Rdsch.* **70** (2), 570–582.

Hinkelbein, K. (1975). Beiträge zur Stratigraphie und Paläontologie des Juras von Ostspanien VIII; Stratigraphie und Fazies im Mitteljura der zentralen Iberischen Ketten. *Neues Jahrb. Geol. Paläontol. Abhandlungen.* **148** (2), 139–184.

Hinz, K., Dostmann, H. and Fritsch, J. (1982). The continental margin of Morocco: seismic sequences, structural elements and geological development. In: Rad, U. von, Hinz, K., Sarnthein, M. and Seibold, E. (eds.) *Geology of the northwest African continental margin.* Springer-Verlag, Berlin, 34–60.

Hirsch, F. (1976). Sur l'origine des particularismes de la faune du Trias et du Jurassique de la plate-forme africano-arabe. *Bull. Soc. géol. Fr. Sér. 7.* **18** (2), 543–552.

Hiss, M. (1982). Neue Ergebnisse zur Paläogeographie des Cenomans in Westfalen. *Neues Jahrb. Geol. Paläont. Monatshefte* **1982** (9), 533–546.

Hobday, D.K., Woodruff, C.M. Jnr. and McBride, M.W. (1981). Paleotopographic and structural controls on non-marine sedimentation of the Lower Cretaceous Antlers Formation and correlatives, North Texas and southeastern Oklahoma. In: Ethridge, F.G. and Flores, R.M. (eds.) *Recent and ancient nonmarine depositional environments: models for exploration. Spec. Publ. Soc. econ. Paleontol. Mineral.* **31**, 71–87.

Hobson, J.P., Fowler, M.L. and Beaumont, E.A. (1982). Depositional and statistical exploration models, Upper Cretaceous offshore sandstone complex, Sussex Member, House Creek Field, Wyoming. *Bull. Am. Assoc. Petrol. Geol.* **66** (6), 689–707.

Hoffmann, D. (1967). Rhat und Lias nordwestlich der Frankenalb auf Blatt Sesslach. *Erlanger geol. Abhandlungen.* **68**, 34pp.

Hoffmann, K. (1949). Zur Paläogeographie des nordwestdeutschen Lias und Doggers. In: Bentz, A. (ed.) *Erdöl und Tektonik in Nordwestdeutschland., Proc. Symp., Hannover, Sept. 1947, Veröff. Amt für Bodenforschung, Hannover-Celle.* 113–129.

Hoffmann, K. (1966). Die Stratigraphie und Paläogeographie der bituminosen Faszies des nordwestdeutschen Oberlias (Toarcium). *Beih. geol. Jahrb.* **58**, 443–498. (Publ. 1968).

Hoffmann, K. (1969). Zur Paläogeographie der Eisenerze des Mittleren und Oberen Dogger in Nordwestdeutschland. In: Bottke, H., Dengler, H., Finkenwirth, A. *et al. Sammelwerk Deutsche Eisenerzlagerstätten: II Eisenerze im Deckgebirge (Postvaristikum) 1. Die Marin-sedimentaren Eisenerze des Jura in Nordwestdeutschland., Beih. geol. Jahrb.* **79**, 204–208.

Hoffmann, K. (1969). Paläogeographie der nordwestdeutschen Lias-Eisenerze. In: Bottke, H., Dengler, H., Finkenwirth, A. *et al. Sammelwerk Deutsche Eisenerzlagerstätten: II Eisenerze im Deckgebirge (Postvaristikum) 1. Die Marin-sedimentaren Eisenerze des Jura in Nordwestdeutschland., Beih. geol. Jahrb.* **79**, 104–110.

Hoffmann , K. (1982). Die Stratigraphie, Paläogeographie und Ammonitenführung des Unter-Pliensbachium (Carixium, Lias gamma) in Nordwest-Deutschland. *Geol. Jahrb.* **A55**, 442pp.

Hoggan, R.D. (1970). Paleontology and paleoecology of the Curtis Formation in the Uinta Mountains Area, Daggett County, Utah. *Brigham Young Univ. Geol. Stud.* **17** (2), 32–58.

Holcomb, C.W. (1971). Hydrocarbon potential of Gulf series of western Gulf Basin. In: Cram, I.H. (ed.) *Future petroleum provinces of the United States – their geology and potential. Mem. Am. Ass. Petrol. Geol.* **15** (2), 887–900.

Holder, H. (1964). Jura. In: Lotze, F. (ed.) *Handbuch der Stratigraphischen Geologie.* 4, F. Enke Verlag, Stuttgart, 603pp.

Hoppe, W. (1976). Zur Bildungsgeschichte des Buntsandsteins in Germanischen Buntsandsteinbecken. *Z. geol. Wiss. D.D.R.* **4** (11), 1457–1471.

Horrath, F. and Channell, J.E.T. (1977). Further evidence relevant to the African/Adriatic promontory as a palaeogeographic premise for Alpine orogeny. In: Biju-Duval, B. and Montadert, L. (eds.) *Structural history of the Mediterranean basins. Proc. Int. Symp., Split, Yugoslavia,* **1976.** Editions Technip, Paris, 133–142.

Horton, A. (1977). The age of the Middle Jurassic 'white sands' of north Oxfordshire. *Proc. Geol. Assoc. Lond.* **88** (3), 147–162.

Houten, F.B. van (1977). Triassic-Liassic deposits of Morocco and eastern North America: comparison. *Bull. Am. Assoc. Petrol. Geol.* **61** (1), 79–99.

Houten, F.B. van (1980). Latest Jurassic – early Cretaceous regressive facies, northeast Africa craton. *Bull. Am. Assoc. Petrol. Geol.* **64** (6), 857–867.

Houten, F.B. van and Brown, R.H. (1977). Latest Paleozoic – early Mesozoic paleography, northwestern Africa. *J. Geol.* **85** (2), 143–156.

Howarth, M.R. (1981). Palaeogeography of the Mesozoic. In: Cocks, L.R.M. (ed.) *The evolving earth, British Museum (Nat. Hist.).* Cambridge Univ. Press, 197–220.

Howell, D.G. and Vedder, J.G. (1978). Late Cretaceous paleogeography of the Salinian block, California. In: Howell, D.G. and McDougall, K.A. (eds.) *Mesozoic paleogeography of the western United States, Proc. 2nd Symp. Pacific Coast Paleogeogr, April 1978, Pacific Sect. Soc. econ. Paleontol. Mineral.* 523–534.

Hoyer, P. (1965). Fazies Paläogeographie und Tektonik des Malm im Deister, Osterewald und Suntel. *Beih. geol. Jahrb.* **61,** 249pp.

Hsu, K.J. (1960). Paleocurrent structures and paleogeography of the Ultrahelvetic Flysch basins, Switzerland. *Bull. geol. Soc. Am.* **71** (5), 577–610.

Hubert, J.F., Butera, J.G. and Rice, R.F. (1972). Sedimentology of Upper Cretaceous Cody-Parkman delta, southwestern Powder River basin, Wyoming. *Bull. geol. Soc. Am.* **83** (6), 1649–1673.

Hubert, J.F. and Hyde, M.G. (1982). Sheet-flow deposits of graded beds and mudstones on an alluvial sandflat-playa system: Upper Triassic Blomidon redbeds, St. Mary's Bay, Nova Scotia. *Sediment. Geol.* **29** (4), 457–474.

Hubert, J.F. and Mertz, K.A. (1980). Eolian dune field of late Triassic age, Fundy Basin, Nova Scotia. *Geology.* **8** (11), 516–519.

Hubert, J.F., Reed, A.A. and Carey, P.J. (1976). Paleogeography of the East Berlin Formation, Newark Group, Connecticut Valley. *Am. J. Sci.* **276** (12), 1183–1207.

Huckriede, R. (1967). Molluskenfaunen mit limnischen und brackischen Elementen aus Jura, Serpulit und Wealden N.W. – Deutschlands und ihre paläogeographische Bedeutung. *Beih. geol. Jahrb.* **67,** 263pp.

Hudson, J.D. (1964). The petrology of the sandstones of the Great Estuarine Series, and the Jurassic palaeogeography of Scotland. *Proc. Geol. Assoc. Lond.* **75** (4), 499–527.

Hunt, A.D. and Ratcliffe, J.D. (1959). Triassic stratigraphy, Peace River area, Alberta and British Columbia, Canada. *Bull. Am. Assoc. Petrol. Geol.* **43** (3), 563–589.

Ibrahim, M.W. (1979). Shifting depositional axes of Iraq: an outline of geosynclinal history. *J. Petrol. Geol.* **2** (2), 181–197.

Ibrahim, M.W. (1981). Lithostratigraphy and subsurface geology of the Albian rocks of South Iraq. *J. Petrol. Geol.* **4** (2), 147–162.

Ibrahim, M.W. (1983). Petroleum geology of Southern Iraq. *Bull. Am. Assoc. Geol.* **67** (1), 97–130.

Imlay, R.W. (1952). Marine origin of Preuss sandstone of Idaho, Wyoming and Utah. *Bull. Am. Assoc. Petrol. Geol.* **36** (9), 1735–1753.

Imlay, R.W. (1957). Paleoecology of Jurassic seas in the Western Interior of the United States. In: Hedgpeth, J.W. and Ladd, H. (eds.) *Treatise on marine ecology and paleoecology, 2, Paleoecology. Mem. geol. Soc. Am.* **67** (2), 469–504.

Imlay, R.W. (1980). Jurassic paleobiogeography of the conterminous United States in its continental setting. *Prof. pap. U.S. geol. Surv.* **1062,** 134pp.

Imlay, R.W. and Detterman, R.L. (1973). Jurassic paleobiogeography of Alaska. *Prof. pap. U.S. geol. Surv.* **801,** 34pp.

Ingersoll, R.V. (1978). Paleogeography and paleotectonics of the Late Mesozoic forearc basin of northern and central California. In: Howell, D.G. and McDougall, K.A. (eds.) *Mesozoic paleogeography of the western United States, Proc. 2nd Symp. Pacific Coast Paleogeogr, April 1978, Pacific Sect. Soc. econ. Paleontol. Mineral.* 471–482.

Ingersoll, R.V. (1979). Evolution of the late Cretaceous forearc basin, northern and central California. *Bull. geol. Soc. Am.* 90 (9), 813–826.

Irving, E.M. (1975). Structural evolution of the northernmost Andes, Colombia. *Prof. pap. U.S. geol. Surv.* **846,** 47pp.

Issawi, B. (1972). Review of Upper Cretaceous-Lower Tertiary stratigraphy in central and southern Egypt. *Bull. Am. Assoc. Petrol. Geol.* **56** (8), 1448–1463.

Jacobshagen, V. (1965). Die Allgäu-Schichten (Jura-Fleckenmergel) zwischen Wettersteingebirge und Rhein. *Jahrb. geol. Bundesanst.* **108,** 1–114.

Jacobshagen, V. (1972). Die Trias der mittleren Ost-Agais und ihre paläogeographischen Beziehungen innerhalb der Helleniden. *Z. Deutsch. geol. Ges.* **123,** 445–454.

Jakobsson, K-H., Hamar, G.P., Ormassen, D.E. and Skarpnes, 0. (1980). Triassic facies in the North Sea, North of the Central highs. In: *Proc. Symp. Sedimentation of the North Sea reservoir rocks, Geilo, Norway May 1980, Nor. Pet. Foren.* **XVIII,** 1–10.

James, G.A. and Wynd, J.G. (1965). Stratigraphic nomenclature of Iranian Oil Consortium Agreement Area. *Bull. Am. Assoc. Petrol. Geol.* **49** (12), 2182-2245.

Jansa, L.F., Bujak, J.P. and Williams, G.L. (1980). Upper Triassic salt deposits of the western North Atlantic. *Can. J. Earth Sci.* **17** (5), 547–559.

Jansa, L.F., Enos, P., Tucholke, B.E., Gradstein, F.M. and Sheridan, R.G. (1979). Mesozoic Cenozoic sedimentary formations of the North American Basin: western North Atlantic. In: Talwani, M., Hay, W. and Ryan, W.B.F. (eds.) *Deep drilling results in the Atlantic Ocean: continental margins and paleoenvironment. Proc. Symp. Harriman, New York, March 1978, Am. geophys. Union, Maurice Ewing Ser.* **3,** 1–57.

Jansa, L.F. and Wade, J.A. (1975a). Geology of the Continental Margin off Nova Scotia and Newfoundland. In: Linden, W.J.M. van der and Wade, J.A. (eds.) *Offshore geology of eastern Canada, 2, Pap. geol. Surv. Can.* **74-30,** 51–106.

Jansa, L.F. and Wade, J.A. (1975b). Palaeogeography and sedimentation in the Mesozoic and Cenozoic south-eastern Canada. In: Yorath, C.J., Parker, E.R. and Glass, D.J. (eds.) *Canada's continental margins and offshore petroleum exploration. Mem. Can. Soc. Petrol. Geol.* **4,** 79–102.

Jaritz, W. (1969-70). Epirogenese in Nordwestdeutschland im höheren Jura und in der Unterkreide. *Geol. Rdsch.* **59** (1), 114–124.

Jaskowiak, M. (1966). Die Oberkreide in Nordpolen. In: Wehrli, H. (ed.) *Proc. Int. Symp. Die Oberkreide der Baltischen Strasse. Ber. Deutsch.*

Ges. geol. Wiss., Ser. A, Geol. Paläeontol. **11** (6), 695–701.

Jaskowiak-Schoeneichowa, M. (1979). Paläogeographie des Alb und Cenoman in Polen. In: Wiedmann, J. (ed.) *Aspekte der Kreide Europas, Proc. 1st Symp. German Cretaceous, Münster, April 1978, Int. Union geol. Sci., Ser. A.* **6**, E. Schweizerbart'sche Verlagsbuchhandlung, Stuttgart, 463–472.

Jaskowiak-Schoeneichowa, M. (1981). Sedymentacja i stratygrafia Kredy Górnej w Potnocno-Zachodniej Polsce. *Pr. Inst. geol.* **98**, 91pp.

Jeletzky, J.A. (1958). Die jüngere Oberkreide (Oberconiacian bis Maastricht) Südwestrusslands und ihr Vergleich mit der Nordweste und Westeuropas. *Beih. geol. Jahrb.* **33**, 157pp.

Jeletzky, J.A. (1971a). Marine Cretaceous biotic provinces and paleogeography of Western and Arctic Canada: illustrated by a detailed study of ammonites. *Pap. geol. Surv. Can.* **70–22**, 92pp.

Jeletzky, J.A. (1971b). Stratigraphy, facies and paleogeography of Mesozoic rocks of Northern and West-Central Yukon. *Pap. geol. Surv. Can.* **71–1**, 203–221.

Jeletzky, J.A. (1972). Stratigraphy facies and paleogeography of Mesozoic and Tertiary rocks of northern Yukon and northwest District of Mackenzie. *Pap. geol. Surv. Can.* **72–1**, 212–215.

Jeletzky, J.A. (1973). Biochronology of the marine boreal latest Jurassic, Berriasian and Valanginian in Canada. In: Casey, R. and Rawson, P.F. (eds.) *The Boreal Lower Cretaceous, Geol. J. Spec. Issue.* **5**, Seel House Press, Liverpool, 41–80.

Jeletzky, J.A. (1974). Contribution to the Jurassic and Cretaceous geology of northern Yukon Territory and District of Mackenzie, northwest Territories. *Pap. geol. Surv. Can.* **74–10**, 23pp.

Jeletzky, J.A. (1975). Jurassic and Lower Cretaceous paleogeography and depositional tectonics of Porcupine plateau, adjacent areas of northern Yukon and those of Mackenzie District. *Pap. geol. Surv. Can.* **74–16**, 52pp.

Jeletzky, J.A. (1976). Mesozoic and ? Tertiary rocks of Quatsino Sound, Vancouver Island, British Columbia. *Bull. geol. Surv. Can.* **242**, 243pp.

Jeletzky, J.A. (1977a). Mid-Cretaceous (Aptian to Coniacian) history of Pacific slope of Canada. In: Kanmera, K. (ed.) *Mid-Cretaceous events. Proc. 2nd Int. Conf., Hokkaido, August-Sept. 1976. Spec. Pap. palaeontol. Soc. Japan.* **21**, 97–126.

Jeletzky, J.A. (1977b). Causes of Cretaceous oscillations of sea level in Western and Arctic Canada and some general geotectonic implications. In: Kanmera, K. (ed.) *Mid-Cretaceous events. Proc. 2nd Int. Con., Hokkaido, August-Sept. 1976. Spec. Pap. palaeontol. Soc. Japan.* **21**, 233–246.

Jeletzky, J.A. (1977c). Porcupine River Formation; a new Upper Jurassic sandstone unit, northern Yukon Territory. *Pap. geol. Surv. Can.* **76–27**, 43pp.

Jeletzky, J.A. (1980a). Dicoelitid belemnites from the Toarcian-Middle Bajocian of western and arctic Canada. *Bull. geol. Surv. Can.* **338**, 38pp.

Jeletzky, J.A. (1980b). Lower Cretaceous and Jurassic rocks of McDougall Pass area and some adjacent areas of north-central Richardson Mountains, northern Yukon Territory and northwestern District of Mackenzie, N.W.T. (NTS-116P/9 and 116P/10): a reappraisal. *Pap. geol. Surv. Can.* **78–22**, 35pp.

Jenny, J., Le Marrec, A. and Monbaron, M. (1981). Les couches rouges du Jurassique Moyen du Haut Atlas central (Maroc):

corrélations lithostratigraphiques, éléments de datations et cadre tectono-sédimentaire. *Bull. Soc. géol. Fr. Sér. 7.* **23** (6), 627–639.

Jensen, 0., Vicente, J.C., Davidson, J. and Goday, E. (1976). Etapas de la evolución marina Jurásica de la Cuenca Andina Externa (mioliminar) entre los paralelos 26° y 29° 30′ sur. In: *Proc. Primer Congr. geol. Chile, Santiago, August 1976, 1, Sect. A, regional and stratigr. geol., Universidad de Chile, Santiago,* **A273-A295**.

Jiazong, F. (1980). The main features of marine Triassic sedimentary facies in southern China. In: Gaetani, M. (ed). *Contributions to the Triassic stratigraphy. Proc. Riccardo Assereto and Giulio Pisa field Symp. on Triassic stratigraphy of the Southern Alps, Bergamo, June 1979. Riv. Ital. Paleontol. Stratigr.* **85** (3–4), 1125–1146.

Jones, D.L. and Irwin, W.P. (1971). Structural implications of an offset early Cretaceous shoreline in northern California. *Bull. geol. Soc. Am.* **82** (4), 815–822.

Jordan, P. (1983). Zur Stratigraphie des Lias zwischen Unterem Hauenstein und Schinznach (Solothurner und Aargauer Faltenjura). *Eclog. geol. Helv.* **76** (2), 355–379.

Juignet, P. (1968). Facies littoraux du Cénomanien des environs du Mans (Sarthe). *Bull. Bur. Rech. géol. minières. Sér. 2, Sect IV,* **4**, 5–20.

Juignet, P., Kennedy, W.J. and Lebert, A. (1978). Le Cénomanien du Maine: formations sédimentaires et faunes d'Ammonites du stratotype. In: Philip, J. (ed.) *Colloque sur le Cénomanien: France-Europe occidentale. Paris, Sept. 1976, Géol. Méditerr., Univ Provence.* **5** (1), 87–100.

Juignet, P., Rioult, M. and Destombes, P. (1973). Boreal influences in the Upper Aptian-Lower Albian beds of Normandy, northwest France. In: Casey, R. and Rawson, P.F. (eds.) *The Boreal Lower Cretaceous, Geol. J. Spec. Issue.* **5**, Seel House Press, Liverpool, 303–326.

Julivert, M. (1973). Les traits structuraux et l'évolution des Andes Colombiennes. *Rev. Géogr. phys. Géol. dyn.* **15** (1–2), 143–156.

Jurgan, H. (1974). Die marine Kalkfolge der Unterkreide in der Quebrada El Way Antofagasta, Chile. *Geol. Rdsch.* **63** (2), 490–516.

Jurgan, H. (1977). Strukturelle und lithofazielle Entwicklung des andinen Unterkreide – Beckens im Norden Chiles (Provinz Atacama). *Geotekton. Forsch.* **52** (1), 1–138.

Jux, U. and Strauch, F. (1968). *Ophiomorpha Lundgren* 1891 aus dem Mesoziokum von Bornholm. *Medd. Dansk geol. Foren.* **18**, 213-219.

Kaever, M. (1983). Aspekte der Kreide Westfalens. *Neues Jahrb. Geol. Paläontol. Abhandlungen.* **166** (1), 86–115.

Kafri, U. (1972). Lithostratigraphy and environment of deposition Judea Group, western and central Galilee, Israel. *Bull. geol. Surv. Israel.* **54**, 56pp.

Kakabadze, M.V. (1979). Facies and Environments of Georgian Ancyloceratids (Lower Cretaceous, USSR). In: Wiedmann, J. (ed.) *Aspekte der Kreide Europas, Proc. 1st Symp. German Cretaceous, Munster, April 1978, Int. Union geol. Sci., Ser. A.* **6**, E. Schweizerbart'sche Verlagsbuchhandlung, Stuttgart, 141–150.

Kamen-Kaye, M. (1982). Mozambique-Madagascar geosyncline, In: Deposition and architecture. *J. Petrol. Geol.* **5** (1), 25–27.

Kaplan, M.Ye. (1974). Distribution of clay minerals in Bajocian and Bathonian sediments of the Arctic region of central Siberia. *Dokl. Acad. Sci. USSR.* **209**, 176–178.

Kaplan, M.Ye. (1975). Composition of Jurassic basal conglomerate and the early Jurassic petrographic provinces of northeastern Siberia. *Dokl. Acad. Sci. USSR.* **218**, 168–170.

Kaplan, M.Ye. and Ronkina, Z.Z. (1972). Distribution of heavy minerals in marine sediments of Mesozoic basins in northern Siberia. *Dokl. Acad. Sci. USSR.* **199**, 173–176.

Kaptsan, V.Kh. and Sokolova, V.V. (1972). Oxfordian reef bodies in the Dobrudja foredeep. *Dokl. Acad. Sci. USSR.* **198**, 84–85.

Katich, P.J. (1953). Source direction of Ferron sandstone in Utah. *Bull. Am. Assoc. Petrol. Geol.* **37** (4), 858–862.

Kauffman, E.G. (1975). Dispersal and biostratigraphic potential of Cretaceous benthonic Bivalvia in the Western Interior. In: Caldwell, W.G.E. (ed.) *The Cretaceous System in the Western Interior of North America. Proc. Int. Symp. Univ. Saskatchewan, Saskatoon, May 1973, Spec. Pap. geol. Assoc. Can.* **13**, 163–194.

Kauffman, G., Kockel, F. and Mollat, H. (1976) Notes on the stratigraphic and paleogeographic position of the Svoula Formation in the Innermost zone of the Hellenides (northern Greece). *Bull. Soc. géol. Fr. Sér 7.* **18** (2), 225–230.

Kaye, P. (1966). Lower Cretaceous palaeogeography of North-West Europe. *Geol. Mag.* **103** (3), 257–262.

Keefer, W.R. (1965). Geologic history of Wind River Basin, central Wyoming. *Bull. Am. Assoc. Petrol. Geol.* **49** (11), 1878–1892.

Kellaway, G.A. and Welch, F.B.A. (1948). *British Regional Geology: Bristol and Gloucester District. 2nd Ed.* H.M.S.O., 91pp.

Kelly, S.R.A. and Rawson, P.F. (1983). Some late Jurassic-mid-Cretaceous sections on the East Midlands Shelf, England, as demonstrated on a field meeting, 18–20 May 1979. *Proc. Geol. Assoc. Lond.* **94** (1), 65–73.

Kemper, E. (1968). Einige Bemerkungen über die Sedimentationsverhältnisse und die fossilen Lebensspuren des Bentheimer Sandsteins (Valanginium). *Geol. Jahrb.* **86**, 49–106.

Kemper, E. (1969). Die Sandsteine der Unterkreide im deutsch-holländischen Grenzgebiet. *Z. Deutsch. geol. Ges.* **119**, 541–545.

Kemper, E. (1973a). The Valanginian and Hauterivian stages in northwest Germany. In: Casey, R. and Rawson, P.F. (eds.) *The Boreal Lower Cretaceous, Geol. J. Spec. Issue.* **5**, Seel House Press, Liverpool, 327–344.

Kemper, E. (1973b). The Aptian and Albian stages in northwest Germany. In: Casey, R. and Rawson, P.F. (eds.) *The Boreal Lower Cretaceous, Geol. J. Spec. Issue.* **5**, Seel House Press, Liverpool, 345–360.

Kemper, E. (1975). Upper Deer Bay Formation (Berriasian–Valanginian) of Sverdrup Basin and biostratigraphy of the Arctic Valanginian. *Pap. geol. Surv. Can.* **75-1B**, 245–254.

Kemper, E. (1979) Die Unterkriede Nordwestdeutschlands Ein Überblick. In: Wiedmann, J. (ed.) *Aspekte der Kreide Europas, Proc. 1st Symp. German Cretaceous, Münster, April 1978, Int. Union geol Sci. Ser. A.* **6**, E. Schweizerbart'sche Verlagsbuchhandlungen, Stuttgart, 1–9.

Kemper, E., Rawson, P.F and Thieuloy, J-P. (1981). Ammonites of the Tethyan ancestry in the early Lower Cretaceous of North-West Europe. *Palaeontol.* **24** (2), 251–311.

Kennerly, J.B. (1980). Outline of the geology of Ecuador. *Overseas Geol. min. Resour. London, N.S.* **55**. H.M.S.O., 19pp.

Kent, H.C. (1966). Paleoenvironments during Conacian time in Colorado. In: Kaufmann, E.G. and Kent, H.C. (eds.) *Paleoenvironments of the Cretaceous seaway in the Western Interior , Prepr. Symp. Colorado School of Mines, May 1967,* Dept Publ. Colorado School of Mines. Golden. 203–236.

Kent, H.C. (1968). Biostratigraphy of Niobrara-equivalent part of Mancos Shale (Cretaceous) in northwestern Colorado. *Bull. Am. Assoc. Petrol. Geol.* **52** (11), 2098–2115.

Kent, P.E. (1968). The Rhaetic Beds. In: Sylvester Bradley, P.C. and Ford, T.D. (eds.) *The Geology of the East Midlands.* Leicester University Press, 174–187.

Kent, P.E. and Perry, J.T.O'B. (1973). The development of the Indian Ocean in Tanzania. In: Blant, G. (ed.) *Sedimentary Basins of the African coasts,* **2**, *south and east coasts, Assoc. Afr. geol Surv.* **113–130**.

Kerr, R.S. (1977) Facies, diagenesis and porosity development in a lower Cretaceous Bank complex, Edwards limestone, north-central Texas. In: Bebout, D.G. and Loucks, R.G. (eds.) *Cretaceous carbonates of Texas and Mexico: application to a subsurface exploration , Rep. Invest. Bur. econ. Geol. Univ. Texas.* **89**, 216–233.

Keupp, H. and Michael, E. (1979). Die Blatterton-Fazies der nordwestdeutschen Unterkreide, Teil 1, Calciodinelloidea aus der Blatterton-Fazies des nordwestdeutschen Unter-Barremium. *Ber. naturhist. Ges.* **122**, 7–69.

Khain, V.E. and Milanovsky, E.E. (1963). Structure tectonique du Caucase d'après les données modernes. In: Durand-Delga, M. (ed.) *Livre à la Mémoire du Professeur Paul Fallot: L'évolution paléogéographique et structurale des domaines méditerranéens et Alpins d'Europe. Mém. Soc. géol. Fr. Hors Sér.l.* **2**, 663–704 .

Khudoley, K.M. and Meyerhoff, A.A. (1971). Palaeogeography and geological history of Greater Antilles. *Mem. geol. Soc. Am.* **129**, 199pp.

Kidd, R.B. and Davies, T.A. (1978). Indian Ocean sediment distribution since the Late Jurassic. *Mar. Geol.* **25** (1), 49–70.

Kieken, M. (1962). Les traits essentiels de la géologie algérienne. In: Durand-Delga, M. (ed.) *Livre à la Mémoire du Professeur Paul Fallot: L'évolution paléogéographique et structurale des domaines méditerranéens et Alpins d'Europe. Mém. Soc. géol. Fr. Hors Sér.* **1**. **1**, 545–614.

Kieken, M. (ed.) (1974). *Atlas géologique du bassin d'Aquitaine.* Publ. Bur. Rech. géol. minières. (Bur. Rech. géol. minières., Soc. Elf-Rech., Soc. Esso-Rech., Soc. natl. Pét. Aquitaine). 28 plates.

Kimura, T. (1974). The ancient continental margins of Japan. In: Burk, C.A. and Drake, C.L. (eds.) *The geology of continental margins.* Springer-Verlag, New York, 817–830.

Kirkaldy, J.F. (1963). The Wealden and marine Lower Cretaceous beds of England. *Proc. Geol. Assoc. Lond.* **74** (2), 127–146.

Kiteley, L.W. (1983). Paleogeography and eustatic-tectonic model of Late Campanian (Cretaceous) sedimentation, southwestern Wyoming and northwestern Colorado. In: Reynolds, M.W. and Dolly, E.D. (eds.) *Mesozoic Paleogeography of the West-Central United States. Proc. 2nd Rocky Mt. Paleogeogr. Symp., Denver, Colorado. Rocky Mt. Sect. Soc. econ. Paleontol. Mineral.* 273–304.

Klein, G. de V. (1963). Regional implications of Triassic paleocurrents, maritime provinces, Canada. *J. Geol.* **71** (6), 801–808.

Klein, V., Muller, V. and Valecka, J. (1979). Lithofazielle und paläeogeographische Entwicklung des Böhmischen Kreidebeckens. In: Wiedmann, J. (ed.) *Aspekte der Kreide Europas, Proc. 1st Symp. German Cretaceous, Münster, April 1978, Int. Union geol. Sci., Ser. A.* **6**, E. Schweizerbart'sche Verlagsbuchhandlung, Stuttgart, 435–446.

Klimetz, M.P. (1983). Speculations on the Mesozoic plate tectonic evolution of eastern China. *Tectonics.* **2** (2), 139–166.

Kluth, C.F. (1983). Geology of the northern Carelo Hills and implications for the Mesozoic tectonics of southeastern Arizona. In: Reynolds, M.W. and Dolly, E.D. (eds.) *Mesozoic Paleogeography of the*

West-Central United States. Proc. 2nd Rocky Mt. Paleogeogr. Symp., Denver, Colorado. Rocky Mt. Sect. Soc. econ. Paleontol. Mineral. 159–172.

Kobayashi, T. (1980). Notes on the Mesozoic history of Thailand and adjacent territories. In: Kobayasni, T., Toriyama, R., Hashimto, N. and Kanno, S. (eds.) *Proc. Symp. Tsukuba, Oct. 1978. Geology and Palaeontology of S.E. Asia.* **21**, 27–36.

Kocurek, G. (1981). Erg reconstruction: the Entrada Sandstone (Jurassic) of northern Utah and Colorado. *Palaeogeogr. Palaeoclimatol. Palaeoecol.* **36**, 125–153.

Kocurek, G. and Dott R.H. Jnr, (1983). Jurassic paleogeography and paleoclimate of the central and southern Rocky Mountain region. In: Reynolds, M.W. and Dolly, E.D. (eds.) *Mesozoic Paleogeography of the West-Central United States. Proc. 2nd Rocky Mt. Paleogeogr. Symp., Denver, Colorado. Rocky Mt. Sect. Soc. econ. Paleontol. Mineral.* 101–118.

Kogbe, C.A. (1973). Geology of the Upper Cretaceous and Tertiary sediments of the Nigerian sector of the Iullemeden basin (West-Africa). *Geol. Rdsch.* **62** (1), 197–211.

Kogbe, C.A. (1976). Palaeogeographic history of Nigeria from Aloian times. In: Kogbe, C.A. (ed.) *Geology of Nigeria. Proc. Conf. Ile-Ife, Nigeria, Jan. 1974.* Elizabethan Publ. Co., Surulere (Lagos), 237–252.

Kogbe, C.A. (1981). Cretaceous and Tertiary of the Iullemeden Basin in Nigeria (West Africa). *Cret. Res.* **2** (2), 129–186.

Kolbel, H. (1967). Die Paläogeographie des Juras im Nordteil der DDR in Beziehung zu den Nachbargebieten. *Ber. Deutsch. Ges. geol. Wiss., A, Geol. Palaeontol.* **12** (3-4), 259–288.

Kolbel H. (1968). Jura. In: Schmidt, K. (ed.) *Grundriss der Geologie der Deutschen Demokratischen Republik, 1, Geologische Entwicklung des Gesamtgebietes.* Akademie Verlag, Berlin, 290-315. (+ Anl.-Abb. 14–16).

Kollmann, H.A. (1978). Les Gastropodes cénomaniens de France et des régions voisines. In: Philip, J. (ed.) *Colloque sur le Cénomanien: France-Europe occidentale. Paris, Sept. 1976, Géol. Méditerr., Univ. Provence.* **5** (1), 101–108.

Konarski, E. (1974). Oksford i kimeryd przedgórza Karpat polskich. *Kwart. geol.* **18** (1), 117–131.

Koop, W.J. and Stoneley, R. (1982). Subsidence history of the Middle East Zagros Basin, Permian to Recent. In: Kent, P., Bott, M.H.P., McKenzie, D.P. and Williams, C.A. (eds.) *The evolution of sedimentary basins. Proc. Symp. R. Soc., June 1981. Philos. Trans. R. Soc . Lond.* **A305** (1489), 149–168 .

Korzh, M.V., Filina, S.I. and Zonn, M.S. (1976). Lithofacies and paleogeographic criteria of distribution of oil and gas pools in the Upper Jurassic sediments of the West Siberian Platform. *Probl. Geol. Nefti . IGiRGI,* **7**, 60-64. [Trans. In: *Petrol. geol.* **15** (6), 268–270. (1978).]

Kotetishvili, E. (1979). Les calcaires urgoniens de Géorgie (U.R.S.S.). In: *L'Urgonien des Pays Méditerranéens. Géobios. Spéc. Mém.* **3**, 193–200 .

Kovacs, S. (1982). Problems of the 'Pannonian Median Massif' and the plate tectonic concept: contributions based on the distribution of Late Paleozoic – early Mesozoic isopic zones. *Geol. Rdsch.* **71** (2), 617–639.

Kraft, J.C., Sheridan, R.E. and Maisano, M. (1971). Time-stratigraphic units and petroleum entrapment models in Baltimore Canyon Basin of Atlantic continental margin geosyncline. *Bull. Am. Assoc. Petrol. Geol.* **55** (5), 658–679.

Krasnov, Ye. V. and Savitskiy, V.O. (1974). Upper Jurassic coral reefs of Sakhalin and the hypothesis of drift of the Japanese islands. *Dokl. Acad. Sci. USSR.* **209**, 52–54.

Kraus, G.P. and Parker, K.A. (1979). Geochemical evaluation of petroleum source rock in Bonaparte gulf – Timor sea region, northwestern Australia. *Bull. Am. Assoc. Petrol. Geol.* **63** (11), 2021–2041.

Kraus, O. (1969). Die Raibler Schichten des Drauzuges (Südliche Kalkalpen). Lithofazielle, sedimentpetrographische und paläogeographische Untersuchungen. *Jahrb. geol. Bundesanst.* **112** (1), 81–152.

Kraus, O. and Ott, E. (1968). Eine ladinische Riff-Fauna im Dobratsch – Gipfelkalk (Kärnten Österreich) und Bemerkungen zum Faziesvergleich von Nordalpen und Drauzug. *Mitt. Bayer. Staatssamml. Paläontol. hist. Geol.* **8**, 263–290.

Kraus, O. and Schmidt-Thome, P. (1967). Faziesverteilung in der alpinen Trias des Heiterwand – Gebietes (Östliche Lechtaler Alpen, Tirol). *Neues Jahrb. Geol. Paläontol. Monatshefte* **1967**, 117–127.

Krimmel, V. (1980). Epirogene Paläotektonik zur Zeit des Keupers (Trias) in Südwestdeutschland. *Arb. Inst. Geol. Paläontol. Univ. Stuttgart, N.F.* **76**, 1–74.

Krishnan, M.S. (1968). *Geology of India and Burma.* Higginbottoms, Madras, 536pp.

Krisl, P. (1969). Die tiefere Sandsteinkeuper in Nordfranken. *Erlanger geol. Abhandlungen.* **75**, 68pp.

Kristan-Tollmann, E. and Tollmann, A. (1982). Die Entwicklung der Tethystrias und Herkunft ihrer Fauna. In: *Proc. Symp. Triassic global geology, Würzburg, Feb. 1982, Geol. Rdsch.* **71** (3), 987–1019.

Kruit, C. and Mandl, G. (1975). Sedimentary 'squeeze structures' indicative of paleoslope in the Upper Cretaceous Flysch Basin of the southern Pyrenees. In: *Primer coloquio de estratIgrafía y paleogeografía del Cretácico de España. Bellaterra-Tremp, November 1973, Publ. Empresa Nac. Adaro. Invest. Mineras, Ser. 7, Trabajos de Congresos y reuniones, Madrid.* **1**, 93–102.

Krumbein, W.C. and Nagel, F.G. (1953). Regional stratigraphic analysis of 'Upper Cretaceous' rocks of Rocky Mountain region. *Bull. Am. Assoc. Petrol. Geol.* **37** (5), 940–960.

Krymholz, G.J. and Louppov, N.P. (1975). Etat de la question sur la limite Jurassique-Crétacé en URSS. In: *Proc. Symp., Colloque sur la limite Jurassique-Crétacé, Lyon, Sept. 1973, Mém. Bur. Rech. géol. minières.* **86**, 350–357.

Krymholz, G.J., Nalnjaeva, T.I. and Sachs, V.N. (1975). L'évolution des Belemnites à la fin du Jurassique et au début du rétacé. In: *Proc. Symp., Colloque sur la limite Jurassique-Crétacé, Lyon, Sept. 1973, Mém. Bur. Rech. géol. minières.* **86**, 237–212.

Krystek, I. and Samuel, O. (1978). Wyskyt kriedy karpatskeho typu severne od Brna (Kurim). *Pr. geol. Krakow.* **71**, 93–110.

Ksiazkiewicz, M. (1960). Pre-orogenic sedimentation in the Carpathian geosyncline. *Geol. Rdsch.* **50**, 8–31.

Ksiazkiewicz, M. (1963). Evolution structurale des Carpathes polonaises. In: Durand-Delga, M. (ed.) *Livre à la mémoire du Professeur Paul Fallot: L'évolution paléogéographique et structurale des domaines méditerranéens et Alpins d'Europe. Mém. Soc. géol. Fr. Hors Sér.* **1, 2**, 529–564.

Kuhn, R. and Dellwig, L.F. (1971). Salt deposits of Permian, Triassic and Tertiary age in W. Germany. In: Muller, G. (ed.) *Sedimentology of parts of Central Europe, Guidebook to excursions. 8th Int. Sedimentol. Congr., Heidelburg, 1971.* Verlag Waldemar Kramer, Frankfurt,

Bibliography

303–320.

Kumar, N. and Gamboa, L.A.P. (1979). Evolution of the Sao Paulo Plateau (southeastern Brazilian Margin) and implications for the early history of the South Atlantic. *Bull. geol. Soc. Am.* **90** (3), 281–293.

Kummel, B. (1957). Paleoecology of Lower Triassic formations of southeastern Idaho and adjacent areas. In: Hedgpeth, J.W. and Ladd, H (eds.) *Treatise on marine ecology and paleoecology, 2, paleoecology. Mem. geol. Soc. Am.* **67** (2), 437–468.

Kummel, B. (1973). Aspects of the Lower Triassic (Scythian) stage. In: Logan, A. and Hills, L.V. (eds.) *The Permian and Triassic systems and their mutual boundary, Proc. Symp. Univ. Calgary, Aug. 1971. Mem. Can. Soc. Petrol. Geol.* **2**, 557–571.

Kunz, I. and Bauss, R. (1975). Über der Bitumenführung und einige Metalle im Posidonienschiefer (Toarc) des Nordwestteils der DDR. *Z. angew. Geol. DDR.* **21** (10), 481–489.

Laing, A.C.M. (1969). Review of geology and case history of petroleum exploration in central Eromanga sub-basin. *J. Aust. Petrol. Explor. Assoc.* **9** (2), 88–96.

Lanteaume, M. (1962). Considérations paléogéographiques sur la partie supposée des Nappes de flysch à Helminthoides des Alpes et des Apennins. *Bull. Soc. géol. Fr. Sér. 7.* **4**, 627–643.

La Torre, L.S. and Villar, J.A. (1970). Paleogeografía del Triásico en el sector occidental de la Cordillera Ibérica. *Estud. geol. Inst. Invest. geol. Lucas Mallada.* **26** (4), 423–430.

Laubacher, G. (1978). Géologie des Andes péruviennes: géologie de la Cordillère orientale et de l'Altiplano au nord et nord-ouest du lac Titicaca (Pérou). *Trav. Doc. O.R.S.T.O.M.* **95**, 217pp

Laubscher, H. and Bernoulli, D. (1977). Mediterranean and Tethys. In: Nairn, A.E.M., Kanes, W.H. and Stehli, F.G. (eds.) *The ocean basins and margins*, **4A**, *The Eastern Mediterranean.* Plenum Press, London, 1–28.

Leckie, G.G. (1982). Lithology and subsidence in the North Sea. In: Kent, P., Bott, M.H.P., McKenzie, D.P. and Williams, C.A. (eds.) *The evolution of sedimentary basins. Proc. Symp. R. Soc., June 1981. Philos. Trans. R. Soc. Lond.* **A305** (1489), 85–100.

Leeder, M.R. and Nami, M. (1979). Sedimentary models for the non-marine Scalby Formation (Middle Jurassic) and evidence for late Bajocian/Bathonian uplift of the Yorkshire basin. *Proc. Yorkshire geol. Soc.* **42** (3), 461–482.

Leeder, M.R. and Zeidan, R. (1977). Giant Late Jurassic sabkhas of Arabian Tethys. *Nature.* **268**, 42–44.

Lefeld, J. (1978). Mongolia. In: Moullade, M. and Nairn, A.E.M. (eds.) *The Phanerozoic geology of the world*, **2**, *The Mesozoic, A.* Elsevier, Amsterdam, 55–78.

LeFranc, J.P. (1976). Etat des connaissances actuelles sur les zonations biostratigraphiques du milieu du Crétacé (Albien à Turonien) au Sahara. In: *Evénements de la partie moyenne du Crétacé: rapports sur la biostratigraphie des régions clés, Uppsala 1975 – Nice 1976, Ann. Mus. Hist. nat. Nice.* **4** XIX, 1–17.

Leggewie, R., Fuchtbauer, H. and El-Najjar, R. (1977). Zur Bilanz des Buntsandsteinbeckens (Korngrobenverteilung und Gesteinsbruchstücke). *Geol. Rdsch.* **66** (2), 551–577.

Le Grand, H.E. (1961). Summary of geology of Atlantic coastal plain. *Bull. Am. Assoc. Petrol. Geol.* **45** (9), 1557–1571.

Le Hegarat, G. (1971). Le Berriasien du sud-est de la France. *Doc. Lab. Géol. Fac. Sci. Lyon.* **43** (2), 309–576.

Lemoine, M. (ed.) (1978). *Geological atlas of Alpine Europe and adjoining Alpine areas.* Elsevier, Amsterdam, 584pp.

Lencinas, A.N. and Salfity, J.A. (1973). Algunas características de la Formación Yacoraite en el oeste de la Cuenca Andina, Provincias de Salta y Jujuy, República Argentina. In: *Proc. 5th Congr. Geológico Argentina, Ciudad de Villa Carlos Paz (Córdoba), October 1972, 3 Sect. 5, Geología Histórica y Estratigrafía, Librart, Buenos Aires. 253–267.*

Leonov, G.P., Loginova, G.A. and Panov, D.I. (1974). Principal zones and regions of sedimentation within the near-middle east part of the Alpine Belt of Europe and Asia in Jurassic, Cretaceous and Palaeogene times. In: *Contributi stratigrafici e paleogeografici sul mesozoico della Tetide. Mem. Riv. Ital. Paleontol. Stratigr.* **14**, 251–289.

Le Pichon, X., Bonnin, J., Francheteau, J. and Sibnet, J-C. (1971). Une Hypothèse d'évolution tectonique du Golfe de Gascogne. In: *Proc. Symp. Histoire structurale du Golfe de Gascogne, Reuil-Malmaison, Déc. 1970, Publ. Inst. Fr. Pét. collection colloq. et séminaires.* **22** (2), Editions Technip, Paris, VI-ll, 1–44.

Le Pichon, X., Melguen, M. and Sibuet, J.C. (1978). A schematic model of the evolution of the South Atlantic. In: Charnock, H. and Deacon, G. (eds.) *Advances in oceanography. Proc. Symp. Joint Oceanogr. assembly, Edinburgh, Sept. 1976.* Plenum Press, 1–48.

Lerand, M. (1973). Beaufort Sea. In: McCrossan, R.G. (ed.) *The future petroleum provinces of Canada – their geology and potential. Mem. Can. Soc. Petrol. Geol.* **1**, 315–386.

Lesta, P.J. and Ferello, R. (1972). Región extra-andina de Chubut y norte de Santa Cruz. In: Leanza, A.F. (ed.) *Geología Regional Argentina, Proc. 1st Symp. Córdoba, Septiembre 1969, Acad. Nac. Cienc., Córdoba.* 601–653.

Lewy, Z. (1972). Recent and Senonian oncolites from Sinai and Southern Israel. *Israel J. earth Sci.* **21**, 193–199.

Lewy, Z. (1975). The geological history of southern Israel and Sinai during the Coniacian. *Israel J. earth Sci.* **24** (1), 19–43.

Lewy, Z. and Raab, M. (1976). Mid-Cretaceous stratigraphy of the Middle East. In: *Evénements de la partie moyenne du Crétacé: rapports sur la biostratigraphie des régions clés, Uppsala 1975 – Nice 1976, Ann. Mus. Hist. nat. Nice.* **4** XXXII, 1–20.

Leyden, R. (1976). Salt distribution and crustal models for the eastern Brazilian margin. In: Almeida, F.F.M. (ed.) *Simposio internacional sobre as margens continentais de tipo Atlántico, Proc. Symp. Sao Paulo, October 1975, An. Acad. bras. Cienc.* **48**, Suppl., 159–168.

Leyden, R., Asmus, H., Zembruscki, S. and Bryan, G. (1976). South Atlantic diapiric structures. *Bull. Am. Assoc. Petrol. Geol.* **60** (2), 196–212.

Libby-French, J. (1981). Lithostratigraphy of Shell 272-1 and 273-1 wells: implications as to depositional history of Baltimore Canyon Trough, Mid-Atlantic OCS. *Bull. Am. Assoc. Petrol. Geol.* **65** (8), 1476–1484.

Lindsey, D.A. (1972). Sedimentary petrology and paleocurrents of the Harebell Formation, Pinyon Conglomerate, and associated coarse clastic deposits, northwestern Wyoming. *Prof. pap. U.S. geol. Surv.* **734-B**, 68pp.

Link, T.A. (1951). Source of oil in 'Tar sands' of Athabaska river, Alberta, Canada. *Bull. Am. Assoc. Petrol. Geol.* **35** (4), 854–8G4.

Lloyd, C.R. (1982). The Mid-Cretaceous earth: paleogeography; ocean circulation and temperature; atmospheric circulation. *J. Geol.* **90** (4), 393–413.

Lock, B.E., Pickton, C.A.G., Smith, D.C., Batten, D.J. and Harland,

W.B. (1978). The geology of Edgeoya and Barentsoya, Svalbard. *Norske Polarinst. Skrifter.* **168**, 64pp.

Longman, M.W. and Mench, P.A. (1978). Diagenesis of Cretaceous limestones in the Edwards aquifer system of south-central Texas: a scanning electron microscope study. *Sediment. Geol.* **21**, 241–276.

López Ramos, E. (1981). *Geología de México*, **3**, 2nd Ed., Inst. Geol. Univ. Nac. México. Edición Escolar, 446pp.

Lorenz, J-F. (1968). La transgression des formations secondaires sur la bordure orientale du Morvan dans la région de Pouilly-en-Auxois et de Saulieu. *Bull. Soc. géol. Fr. Sér. 7.* **10**, 282–286.

Lott, G.K., Knox, R.W.O'B., Bigg, P.J., Davey, R.J. and Morton, A.C. (1980). Aptian-cenomanian stratigraphy in boreholes from offshore south-west England. *Rep. Inst. geol. Sci. Lond.* **80** (8), 12pp.

Loucks, R.G. (1977). Porosity development and distribution in shoal – water carbonate complexes – subsurface Pearsall Formation (Lower Cretaceous) South Texas. In: Bebout, D.G. and Loucks, R.G. (eds.) *Cretaceous carbonates of Texas and Mexico: applications to subsurface exploration, Rep. Invest. Bur. econ. Geol. Univ. Texas.* **89**, 97–126.

Loughman, D.L. and Hallam, A. (1982). A facies analysis of the Pucara Group (Norian to Toarcian carbonates, organic-rich shale and phosphate) of Central and Northern Peru. *Sediment. Geol.* **32** (3), 161–194.

Love, J.D. (1973). Harebell Formation (Upper Cretaceous) and Pinyon Conglomerate (Uppermost Cretaceous and Paleocene), northwestern Wyoming. *Prof. pap. U.S. geol. Surv.* **734-A**, 54pp.

Lowe, D.R. (1972). Implications of three submarine mass movement deposits, Cretaceous, Sacramento Valley, California. *J. sediment. Petrol.* **42** (1), 89–101.

Lowrey, R.O. (1976). Paleoenvironment of the Carmel Formation at Sheep Creek Gap, Daggett County, Utah. *Brigham Young Univ. Geol. Stud.* **23** (1), 173–202.

Ludbrook, N.H. (1971). Stratigraphy and correlation of marine sediments in the western part of the Gambier Embayment. In: Wopfner, H. and Douglas, J.G. (eds.) *The Otway Basin of southeastern Australia. Spec. Bull. geol. Surv. South Aust. Victoria*, 47–66.

Ludbrook, N.H. (1978). Australia. In: Moullade, M. and Nairn, A.E.M. (eds.) *The Phanerozoic geology of the world*, **2**, *The Mesozoic, A.* Elsevier, Amsterdam, 209–250.

Macdonald, D.I.M. and Tanner, P.W.G. (1983). Sediment dispersal patterns in part of a deformed Mesozoic back-arc basin on South Georgia, South Atlantic. *J. Sediment. Petrol.* **53** (1), 83–104.

McGookey, D.P., Haun, J.D., Hale, L.A., Goodell, H.G., McCubbin, D.G., Weimer, R.J. and Wulf, G.R. (1972). Cretaceous system. In: *Geologic Atlas of the Rocky Mountains Region, U.S.A., Rocky Mt. Assoc. Geol.* 190–228.

McGowen, H., Granata, G.E. and Seni, S.J. (1983). Depositional setting of the Triassic Dockum Group, Texas Panhandle and eastern New Mexico. In: Reynolds, M.W. and Dolly, E.D. (eds.) *Mesozoic paleogeography of the West-Central United States, proc. 2nd Rocky Mountain Paleogeogr. Symp., Denver, Colorado. Rocky Mt. Sect., Soc. econ. Paleontol. Mineral.* 13–38.

McGregor, A.A. and Biggs, C.A. (1968). Bell Creek field, Montana: a rich stratigraphic trap. *Bull. Am. Assoc. Petrol. Geol.* **52** (10), 1869–1887.

MacLachlan, M.M. (1972). Triassic system. In: *Geologic Atlas of the Rocky Mountains Region, U.S.A., Rocky Mt. Assoc. Geol.* 166–176.

McLane, M. (1982). Upper Cretaceous coastal deposits in south-central Colorado-Codell and Juana Lopez Members of Carlile Shale. *Bull. Am. Assoc. Petrol. Geol.* **66** (1), 71–90.

McLean, J.R. (1979). Regional considerations of the Elmworth Field and the Deep Basin. *Bull. Can. Pet. Geol.* **27** (1), 53–62.

McLean, J.R. and Wall, J.H. (1981). The Early Cretaceous Moosebar Sea in Alberta. *Bull. Can. Pet. Geol.* **29** (3), 334-377.

McMannis, W.J. (1965). Resumé of depositional and structural history of western Montana. *Bull. Am. Assoc. Petrol. Geol.* **49** (11), 1801–1823.

McWhae, J.R.H., Elie, R., Laughton, K.C. and Gunther, P.R. (1980). Stratigraphy and petroleum prospects of the Labrador Shelf. *Bull. Can. Pet. Geol.* **28** (4), 460–488.

Maack, V. (1981). Zur paläogeographischen Rekonstruktion erdgeschichtlicher Ozeane. In: *Proc. Internationalen Alfred Wegener Symp., Berlin Feb. 1980. Geol. Rdsch.* **70** (1), 78–90.

Mabesoone, J.M. and Tinoco, I.M. (1973). Palaeoecology of the Aptian Santana Formation (northeastern Brazil). *Palaeogeogr. Palaeoclimatol. Palaeoecol.* **14** (2), 97–118.

Mabesoone, J.M., Tinoco, I.M. and Coutinho, P.N. (1968). The Mesozoic-Tertiary boundary in northeastern Brazil. *Palaeogeogr. Palaeoclimatol. Palaeoecol.* **4** (3), 161–185.

Machens, E. (1973). The geologic history of the marginal basins along the northshore of the Gulf of Guinea. In: Nairn, A.E.M. and Stehli, F.G. (eds.) *The ocean basins and margins*, **1**, *The South Atlantic*. Plenum Press, London, 351–390.

Mader, D. (1982). Aeolian sands in continental red beds of the Middle Buntsandstein (Lower Triassic) at the western margin of the German Basin. *Sediment. Geol.* **31** (3), 191–230.

Mader, D. (1984). Evolution of fluvial sedimentation in the Buntsandstein (Lower Triassic) of the Eifel (Germany). *Sediment. Geol.* **37** (1–2), 1–84.

Mahjoubi, H. and Samama, J-C. (1983). Modèle de concentration supergène d'amas sidéritique: le cas du Jebel Jerrissa, Tunisie. *Bull. Soc. géol. Fr. Sér. 7.* **25** (1), 91–99.

Maillard, J. (1965). Le passage Barremien-Aptien et ses rapports avec l'Urgonien dans le couloir rhodanien entre Valence et Avignon. In: *Colloque sur le Crétacé Inférieur, Proc. Int. Symp. Lyon, Sept. 1963, Mém. Bur. Rech. géol. minières.* **34**, 147–156.

Malinowska, L. (1976). Boreal faunal influences in the Lower and Middle Oxfordian of Poland. *Biul. Inst. geol.* **291**, 5–40.

Malinowska, L. (1980). Ammonites of the Genus *Popanites*, Rollier, 1909, from the Lower Oxfordian in Poland. *Biul. Inst. geol.* **323**, 29–39.

Malinowska, L. (1981). The *Cawtoniceras tenuiserratum* zone of the Middle Oxfordian of Poland. *Biul. Inst. geol.* **335**, 27–50.

Malumian, N., Nullo, F.E. and Ramos, V.A. (1983). The Cretaceous of Argentina, Chile, Paraguay and Uruguay. In: Moullade, M. and Nairn, A.E.M. (eds.) *The phanerozoic geology of the world*, **2**, *the Mesozoic, B.* Elsevier, Amsterdam, 265–304.

Mangin, J-P. and Rat, P. (1962). L'évolution post-hercynienne entre Asturies et Aragon (Espagne). In: Durand-Delga, M. (ed.) *Livre à la Mémoire du Professeur Paul Fallot: L'évolution paléogéographique et structurale des domaines méditerranéens et Alpins d'Europe. Mém. Soc. géol. Fr. Hors Sér.1*, **1**, 333–350.

Mangold, C., Enay, R. and Dominjon, P. (1964). Note préliminaire sur le Bathonien inférieur du Bugey (Jura méridional). *Bull. Soc. géol.*

Fr. Sér. 7. **6** (4), 529–534.

Manspeizer, W. (1982). Triassic-Liassic basins and climate of the Atlantic passive margins. In: *Proc. Symp. Triassic global geology, Wurzburg, Feb. 1982. Geol. Rdsch.* **71** (3), 895–917.

Manspeizer, W., Puffer, J.H. and Cousminer, H.L. (1978). Separation of Morocco and eastern North America: a Triassic-Liassic stratigraphic record. *Bull. geol. Soc. Am.* **89** (6), 901–920.

Marchand, D. and Thierry, J. (1974). Les influences Mésogéennes et boréales dans le Callovien de Bourgogne. *Bull. Soc. géol. Fr. Sér. 7.* **16** (4), 476–489.

Marcher, M.V. and Stearns, R.G. (1962). Tuscaloosa Formation in Tennessee. *Bull. geol. Soc. Am.* **73** (11), 1365–1386.

Marchese, H.G. (1971). Litoestratigrafía y variaciones faciales de las sedimentitas mesozoicas de la Cuenca Neuquina, Prov. de Neuquen, Rep. Argentina. *Rev. Assoc. geol. Argentina.* **26** (3), 343–410.

Marek, S. (1968). Zarys stratygrafii kredy dolnej niecki brzeznej. *Kwart. geol.* **12** (2), 345–368.

Marek, S. (1977). Budowa geologiczna wschodniej części niecki Mogilensko-Łódzkiej (Strefa Gopto-Ponetow-Pabianice). *Pr. Inst. Geol. Warszawa.* **80**, 165pp.

Marek, S. (1983). Epikontinentale Unterkreide-Ablagerungen in Polen. In: *2nd Symp. Kreide, Munich 1982. Zitteliana.* **10**, 55–64.

Marek, S., Bielecka, W. and Sztejn, J. (1969). Górny portland (wolg) i berias (riazan) na Niżu Polskim. *Kwart. geol.* **13** (3), 566–582.

Marek, S. and Raczynska, A. (1973). The stratigraphy and palaeogeography of the Lower Cretaceous deposits of the Polish lowland area. In: Casey, R. and Rawson, P.F. (eds.) *The Boreal Lower Cretaceous, Geol. J. Spec. Issue.* **5**, Seel House Press, Liverpool, 369–386.

Marek, S. and Raczynska, A. (1979). Paläogeographie der Unterkreide des nordpolnischen Beckens. In: Wiedmann, J. (ed.) *Aspekte der Kreide Europas, Proc. 1st Symp. German Cretaceous, Münster, April 1978, Int. Union geol. Sci., Ser. A.* **6**, E. Schweizerbart'sche Verlagsbuchhandlung, Stuttgart, 447–462.

Maresch, W.V. (1983). The northern Andes. In: Moullade, M. and Nairn, A.E.M. (eds.) *The Phanerozoic geology of the world,* **2,** *the Mesozoic, B.* Elsevier, Amsterdam, 121–149.

Marie, J., Trouve, P., Desforges, G. and Dufaure, P. (1982). Nouveaux éléments de paléogéographie du Crétacé de Tunisie. In: Cotillon, P. and Phillip, J. (eds.) *Proc. Mtg. Les Facies de plate-forme du Crétacé dans le domaine Mésogéen, 26th Int. geol. Congr., Paris, 1980, Cret. Res.* **3** (1–2), 167–170.

Markovskiy, N.I. (1975). Paleogeographic research in study of oil-gas basins. *Int. geol. Rev.* **17** (7), 757–761.

Markovskiy, N.I. (1981). Influence of an ancient river network on formation of accumulations of oil and gas in western Siberia. *Int. geol. Rev.* **23** (7), 781–784.

Marocco, R. (1978). Géologie des Andes Péruviennes: un segment E-W de la chaîne des Andes péruviennes, la déflexion d'Abancay. Etude géologique de la Cordillère orientale et des hauts plateaux entre Cuzco et San Miguel sud du Pérou (12° 30' S a 14° 00' S). *Trav. Doc. O.R.S.T.O.M.* **94**, 195pp.

Marsaglia, K.M. and De Vries Klein, G. (1983). The paleogeography of Paleozoic and Mesozoic storm depositional systems. *J. Geol.* **91** (2), 117–142.

Martin, H. (1982). Die Trias im Südlichen Afrika. In: *Proc. Symp. Triassic global geology, Wurzburg, Feb. 1982. Geol. Rdsch.* **71** (3), 937–947.

Martin, L.J. (1959). Stratigraphy and depositional tectonics of North Yukon-Lower Mackenzie area, Canada. *Bull. Am. Assoc. Petrol. Geol.* **43** (10), 2399–2455.

Martínez, C. (1980). Géologie des Andes Boliviennes: Structure et évolution de la chaîne hercynienne et de la chaîne andine dans le nord de la Cordillère des Andes de Bolivie. *Trav. Doc. O.R.S.T.O.M.* **119**, 353pp.

Marzolf, J.E. (1982). Paleogeographic implications of Early Jurassic (?) Navajo and Aztec sandstones. In: Frost, E.G. and Martin, D.L. (eds.) *Mesozoic-Cenozoic tectonic evolution of the Colorado River Region, California, Arizona and Nevada. (Anderson-Hamilton Volume).* Cordilleran, San Diego, 493–501.

Marzolf, J.E. (1983). Changing wind and hydrologic regimes during deposition of the Navajo and Aztec Sandstones; Jurassic (?) southwestern U.S. In: Brookfield, M.E. and Ahlbrandt, T.S. (eds.) *Eolian sediments and processes. Devs. in Sedimentol.* **38**, Elsevier, 635–60.

Masetti, D. and Neri, C. (1980). L'Anisico della Val di Fussa (Dolomiti Occidentali): sedimentologia e paleogeografia *Ann. Univ. Ferrara N.S., Ser. 9, Sci. geol. paleontol.* **7** (1), 19pp.

Massaad, M. (1976). Origin and environment of deposition of Lebanon Basal Sandstones. *Eclog. geol. Helv.* **69** (1), 85–91.

Massari, F., Medizza, F. and Channell, J.E.T. (1983). Santonian to Maastrichtian stratigraphy of some pelagic limestone sections of the Venetian Alps (Northern Italy). *Newsl. Stratig.* **12** (1), 18–28.

Masse, J-P. and Allemann, J. (1982). Relations entre les séries carbonatées de plate-forme Provencale et Sarde au Crétacé Inférieur. In: Cotillon, P. and Phillip, J. (eds.) *Proc. Mtg. Les facies de plate-forme du Crétacé dans le domaine Mésogéen, 26th Int. geol. Congr., Paris, 1980, Cret. Res.* **3** (1–2), 19–34.

Masse, J-P., Giroud-d'Argoud, G., Thieuloy, J-P. and Tronchetti, G. (1980). Le Calcaire de Fontdouille (Clansayesien p.p.) en Basse Provence méridionale: stratigraphie et sédimentologie. Ses relations avec les formations clansayesiennes et la paléographie du sud-est de la France. *Géol. Méditerr., Univ. Provence.* **7** (3), 261–276.

Masse, J-P. and Philip, J. (1981). Cretaceous coral-rudist buildups of France. In: Toomey, D.F. (ed.) *European fossil reef models. Spec. Publ. Soc. econ. Paleontol. Mineral.* **30**, 399–426.

Masse, J-P. and Thieuloy, J-P. (1979). Précisions sur l'âge des calcaires et des formations associées de l'Aptien sud-constantinois (Algérie). Conséquences paléogéographiques. *Bull. Soc. géol. Fr. Sér. 7.* **21** (1), 65–71.

Masson, D.G. and Roberts, D.G. (1981). Late Jurassic-Early Cretaceous reef trends on the continental margin S.W. of the British Isles. *J. geol. Soc. Lond.* **138** (4), 437–443.

Mathieu, M. (1965). Le Néocomien dans le Sud du Bassin de Paris. In: *Proc. Colloque sur le Crétacé Inférieur, Lyon, Sept. 1963, Mém. Bur. Rech. géol. minières.* **34**, 587–604.

Matsumoto, T. (1978). Japan and adjoining areas. In: Moullade, M. and Nairn, A.E.M. (eds.) *The Phanerozoic geology of the world,* **2,** *The Mesozoic, A.* Elsevier, Amsterdam, 79–144.

Mattauer, M. and Seguret, M. (1971). Les relations entre la chaîne des Pyrénées et le Golfe de Gascogne. In: *Proc. Symp. Histoire structurale du Golfe de Gascogne, Rueil-Malmaison, Déc. 1970, Publ. Inst. Fr. Pét., collection colloq. et séminaires.* **21** (1), Editions Technip, Paris, IV-4, 1–24.

Megard, F. (1979). Estudio geológico de los Andes del Perú central. *Bol. Inst. geol. Min. Metal. Lima., Ser D. Estud. Esp.* **8**, 227pp.

Megard-Galli, J. and Baud, A. (1977). Les milieux carbonatés du Trias et l'application de méthodes sédimentologiques comme outil de corrélation (France et régions limitrophes). In: *Proc. Colloq. Trias Francais, Orléans, Jan. 1977. Bull. Bur. Rech. géol. minières., ser. 2, sect. IV,* **3**, 233–250.

Megelinik-Assenat, S. (1983). Le Lias inférieur de la bordure Aquitaine du Massif Central Francais: premières manifestations de l'ouverture de l'Atlantique. *Bull. Cent. Rech. Explor. Prod. Elf-Aquitaine.* **7** (1), 45–67.

Megnien, C. (1980) Tectogénèse du Bassin de Paris: étapes de l'évolution du bassin. *Bull. Soc. géol. Fr. Sér.* 7. **22** (4), 669–680.

Megnien, C., Megnien, F. and Debrand-Passard, S. (eds.) (1980). Synthèse géologique du bassin de Paris: 1. Stratigraphie et paléogéographie. *Mém. Bur. Rech. géol. minières.* 101, 466pp.

Megnien, C., Megnien, F. and Debrand-Passard, S. (eds.) (1980). Synthèse géologique du bassin de Paris: 2. Atlas. *Mém. Bur. Rech. géol. minières.* **102**, 55 maps.

Megnien, C., Megnien, F. and Debrand-Passard, S. (eds.) (1980). Synthèse géologique du bassin de Paris: 3. Lexique des noms de Formation. *Mém. Bur. Rech. géol. minières.* **103**, 466pp.

Meiburg, P. (1979). Kreide-Transgression und Palaokarst im Gebiet der Warsteiner Karbonat-Platform (Westfalen). In: Wiedmann, J. (ed.) *Aspekte der Kreide Europas, Proc. 1st Symp. German Cretaceous, Münster, April 1978, Int. Union geol. Sci., Ser. A.* **6**, E. Schweizerbart'sche Verlagsbuchhandlung, Stuttgart, 363–384.

Meister, E.M. and Aurich, N. (1972). Geologic outline and oil fields of Sergipe basin, Brazil. *Bull. Am. Assoc. Petrol. Geol.* **56** (6), 1034–1047.

Melguen, M., Le Pichon, X. and Sibuet, J.C. (1978). Paléoenvironnement de l'Atlantique sud. *Bull. Soc. géol. Fr. Sér.* 7. **20** (4), 471–489.

Mello, J. (1974). Facial development and facial relations of the Slovak Karst Middle and Upper Triassic (West Carpathians, Southern part of Germerids). In: Zapfe, H. (ed.) *Die Stratigraphie der alpin-mediterranean Trias, Proc. Symp. Wien, May 1973.* Springer-Verlag, Wien, 147–156.

Mello, J. (1976). Facialne a stratigraficke clenenie wettersteinskych vapencov cs. uzemia planiny Dolny vrch. *Geol. Prace. Spravy.* **65** (1), 37–52.

Mello, J. (1977). Aplikacia litofacialnej analyzy triasovych vapencov v tektonike (juzna cast plesivkey planiny). *Geol. Prace, Spravy.* 67, 175–180.

Mello, J. and Polak, M. (1978). Facial and paleogeographical outline of the West Carpathians Middle Triassic. Illyrian-Langobardian. In: Vozar, J. (ed.) *Paleogeographical evolution of the West Carpathians, Proc. Symp. Bratislava, April 1977, Geologicky, ustav Dionyza Stura, Bratislava.* 301–311.

Mello, J.F. (1969). Foraminifera and stratigraphy of the upper part of the Pierre Shale and lower part of the Fox Hills Sandstone (Cretaceous) north-central South Dakota. *Prof. pap. U.S. geol. Surv.* **611**, 121pp.

Mencher, E., Fichter, H.J., Renz, H.H., Wallis, W.E., Renz, H.H., Patterson, J.M. and Robie, R.H. (1953). Geology of Venezuela and its oil fields. *Bull. Am. Assoc. Petrol. Geol.* **37** (4), 690–777.

Mendieta, C.C. (1977). El Sistema Cretácico en la parte Boliviana de la Cuenca Cretácica Andina. *Rev. Tec. Y.P.F.B.* **6** (1–2), 5–46.

Mensink, H. (1965). Stratigraphie und Paläogeographie des marinen Jura in den nordwestlichen Iberischen Ketten (Spanien). In: Beuther, A., Dahm, H., Kneuper-Haack, F. *et al. Der Jura und Wealden in Nordost-Spanien. Beih. geol. Jahrb.* **44**, 55–102 (1966).

Mensink, H. and Schudack, M. (1982). Caliche, Bodenbildungen und die paläogeographische Entwicklung an der Wendemariner Jura/Wealden in der Westlichen Sierra de los Cameros (Spanien). *Neues Jahrb. Geol. Paläontol. Abhandlungen.* **163** (1), 49–80.

Merabet, O. and Popov, A. (1972). Les bassins salifères de l'Algérie. In: Richter-Bernburg, G. (ed.) *Geology of saline deposits, Proc. Symp. Int. Union geol. Sci., Hanover, May 1968, UNESCO, Paris.* 173–181.

Mercado, M. (1981). Jurassic-Neocomian paleogeography of the Atacama Region, Chile. In: Miller, H. and Rosenfeld, U. (eds.) *Beitrage zum VII Geowissenschaftlichen Lateinamerika-Kolloquium, Heidelberg 1980, Zentralbl. Geol. Paläontol.* **1** (3–4), 356–358.

Merriam, D.F. (1955). Jurassic rocks in Kansas. *Bull. Am. Assoc. Petrol. Geol.* **39** (1), 31–46.

Mesner, J.C. and Wooldridge, L.C.P. (1964). Maranhao Paleozoic basin and Cretaceous coastal basins, north Brazil. *Bull. Am. Assoc. Petrol. Geol.* **48** (9), 1475–1512.

Meyer, K. (1969). Zur Palaeogeographie der Stufen Rhat bis Alb im Emsland. *Erdöl Erdgas Z.* **85** (12), 484–505.

Meyer, R. (1981). Jura nördlich der Alpen. 4.3. Malm (Weisser oder Uberer Jura). In: Haunschild, H. and Jerz, J. (eds.) *Erläuterungen zur Geologischen Karte von Bayern 1:500,000.* Bayerisches Geologisches Landesamt, München. 62–68.

Meyer, R. (1981). Kreide nördlich der Alpen. In: Haunschild, H. and Jerz, H. (eds.) *Erläuterungen zur Geologischen Karte von Bayern 1:500,000.* Bayerisches Geologisches Landesamt, München. 68–70.

Meyer, R. and Schmidt-Kaler, H. (1981). Jura nördlich der Alpen. 4.1. Lias (Schwarzer oder Unterer Jura). In: Haunschild, H. and Jerz, H. (eds.) *Erläuterungen zur Geologischen Karte von Bayern 1:500,000.* Bayerisches Geologisches Landesamt, München. 55–57.

Meyer, R. and Schmidt-Kaler, H. (1981). Jura nördlich der Alpen. 4.2. Dogger (Brauner oder Mittlerer Jura). In: Haunschild, H. and Jerz, H. (eds.) *Erläuterungen zur Geologischen Karte von Bayern 1:500,000.* Bayerisches Geologisches Landesamt, München. 58–61.

Meyer, R.K.F. (1972). Stratigraphie und Fazies des Frankendolomits (Malm). 1. Teil: Nördliche Frankenalb. *Erlanger geol. Abhandlungen.* **91**, 28pp.

Meyer, R.K.F. (1972). Stratigraphie und Fazies des Frankendolomits (Malm). 2 Teil: Mittlere Frankenalb. *Erlanger geol. Abhandlungen.* **96**, 34pp.

Meyer, R.K.F. (1977). Stratigraphie und Fazies des Frankendolomits und der Massenkalke (Malm). 3, Teil: Südliche Frankenalb. *Erlanger geol. Abhandlungen.* **104**, 40pp.

Meyer, R.K.F. (1981). Die Küste des Obercenoman; Meeres (Oberkreide) westlich von Amberg. *Geol. Bl. N.O. Bayern.* **31** (1–4), 306–21.

Meyerhoff, A.A. (1970a). Continental Drift: implications of palaeomagnetic studies, meteorology, physical oceanography and climatology. *J. Geol.* **78** (1), 1–51.

Meyerhoff, A.A. (1970b). Continental Drift, II: high-latitude evaporite deposits and geologic history of Arctic and North Atlantic oceans. *J. Geol.* **78** (5), 406–444.

Meyerhoff, A.A. (1973). Mass biotal extinctions, world climate changes and galactic motions: possible inter-relations. In: Logan, A. and Hills, L.V. (eds.) The Permian and Triassic systems and their mutual boundary, *Proc. Symp. Univ. Calgary, Aug. 1971, Mem. Can. Soc. Petrol. Geol.* **2**, 745–758.

Meyerhoff, A.A. and Hatten, C.W. (1974). Bahamas salient of North America. In: Burk, C.A. and Drake, C.L. (eds.) *The geology of continental margins.* Springer-Verlag, New York, 429–446.

Meyerhoff, A.A. and Hatten, C.W. (1974). Bahamas salient of North America: tectonic framework, stratigraphy and petroleum potential. *Bull. Am. Assoc. Petrol. Geol.* **58** (6, II), 1201–1239.

Meyerhoff, A.A., Khudoley, K.M. and Hatten, C.W. (1969). Geologic significance of radiometric dates from Cuba. *Bull. Am. Assoc. Petrol. Geol.* **53** (12), 2494–2500.

Miall, A.D. (1973). Regional geology of northern Yukon. *Bull. Can. Pet. Geol.* **21** (1), 81–116.

Miall, A.D. (1976). Palaeocurrent and palaeohydrologic analysis of some vertical profiles through a Cretaceous braided stream deposit, Banks Island, Arctic Canada. *Sedimentol.* **23** (4), 459–483.

Miall, A.D. (1979). Mesozoic and Tertiary geology of Banks Island, Arctic Canada: the history of an unstable craton margin. *Mem. geol. Surv. Can.* **387**, 235pp.

Miall, A.D. (1981). Late Cretaceous and Paleogene sedimentation and tectonics in the Canadian Arctic Islands. In: Miall, A.D. (ed.) *Sedimentation and tectonics in alluvial basins, Spec. Pap. geol. Assoc. Can.* **23**, 221–272.

Michael, E. (1974). Zur Paläokologie und Faunenführung im westlichen Bereich des norddeutschen Unterkreide-Meeres. *Geol. Jahrb.* **A19**, 68pp.

Michael, E. (1979). Mediterrane Fauneneinflüsse in den borealen Unterkreide-Becken Europas, besonders Nordwestdeutschlands. In: Wiedmann, J. (ed.) *Aspekte der Kreide Europas, Proc. 1st Symp. German Cretaceous, Münster, April 1978, Int. Union geol. Sci., Ser. A.* **6**, E. Schweizerbart'sche Verlagsbuchhandlung, Stuttgart, 305–322.

Michalik, J. (1973). Paläogeographische Studie des Rats der Krizna-Decke des Strazov-Gebirges und einiger anliegender Gebiete. *Geol. Sb. Bratislava.* **24** (1), 123–140.

Michalik, J. (1974). Zur Paläogeographie der ratischen Stufe des westlichen Teils der Krizna-Decke in den Westkarpaten. *Geol. Sb. Bratislava.* **25** (2), 257–285.

Michalik, J. (1977). Paläogeographische Untersuchungen der Fatra-Schichten (Kossen-Formation) des Nördlichen Teiles des Fatrikums in den Westkarpaten. *Geol. Sb. Bratislava.* **28** (1), 71–94.

Michalik, J. (1978). To the paleogeographic, paleotectonic and paleoclimatic development of the West Carpathian area in the Uppermost Triassic. In: Vozar, J. (ed.) *Paleogeographical evolution of the West Carpathians, Proc. Symp. Bratislava, April 1977, Geologicky ustav Dionyza Stura, Bratislava.* 189–209.

Michalik, J. (1980). A paleoenvironmental and paleoecological analysis of the West Carpathian part of the northern Tethyan nearshore region in the Latest Triassic time. In: Gaetani, M. (ed.) *Contributions to the Triassic stratigraphy. Proc. Riccardo Assereto and Giulio Pisa field Symp. on Triassic stratigraphy of the southern Alps, Bergamo, June 1979. Riv. Ital. Paleontol. Stratigr.* **85** (3–4), 1047–1064.

Michalik, J. and Kovac, M. (1982). On some problems of palinspastic reconstructions and Ceno-Mesozoic paleogeographical development of the Western Carpathians. *Geol. Sb. Bratislava.* **33** (4), 481–507.

Michard, A., Aubague, M., Lefavrais-Raymond, A. and L'Homer, A. (1979). Le Lotharingien supérieur dans le bassin des Causses: stratigraphie et évolution du bassin. *Bull. Soc. géol. Fr. Sér. 7.* **21** (1), 3–10.

Michard, A.G. and Coumol, A. (1978). Sédimentation liassique dans les Causses: contrôle des minéralisations Zn-Pb associées au Lotharingien. *Bull. Bur. Rech. géol. minières. sér. 2, sect. II,* **2**, 57–120.

Michelsen, O. (1975). Lower Jurassic biostratigraphy and ostracods of the Danish embayment. *Dan. geol. unders. Ser. 2* **104**, 1–287.

Michelsen, O. (1978). Stratigraphy and distribution of Jurassic deposits of the Norwegian-Danish Basin. *Dan. geol. unders. Ser. B.* **2**, 28pp.

Middlemiss, F.A. (1962a). Brachiopods and shorelines in the Lower Cretaceous. *Ann. Mag. nat. Hist. ser. 13.* **4**, 613–626.

Middlemiss, F.A. (1962b). Brachiopod ecology and Lower Greensand palaeogeography. *Palaeontol.* **5** (2), 253–267.

Middlemiss, F.A. (1975). Studies in the sedimentation of the Lower Greensand of the Weald, 1875-1975: a review and commentary. *Proc. Geol. Assoc. Lond.* **86** (4), 457–473.

Miladinovic, M. (1976). Vue des relations paléogéographiques du Mésozoïque de la Bosnie orientale. *Bull. Geol. Inst. Rech. Geol. Titograd.* **8**, 99–110.

Milewicz, J. (1979). Rozmieszczenie osadów kredowych w basenie potnocnosudeckim. *Kwart. geol.* **23** (4), 819–826.

Mills, R.A., Hugh, K.E. Feray, D.E. and Swolfs, H.C. (1967). Mesozoic stratigraphy of Honduras. *Bull. Am. Assoc. Petrol. Geol.* **51** (9), 1711–1786.

Minato, M., Gorai, M. and Hunahashi, M. (eds.) (1965). *The geologic development of the Japanese Islands.* Tsukiji Shokan, Tokyo, 442pp.

Misik, M. (1974). Paleogeographic outline of the Tithonian in the Czechoslovakian Carpathians. *Acta geol. Pol.* **24** (3), 485–503.

Misik, M. (1978). Kontinentalne, brakicke a hypersalinicke facie v mezozoiku centralnych Zapadnych Karpat a otazka vynorenych oblasti. In: Vozar, J. (ed.) *Paleogeographic evolution of the West Carpathians, Proc. Symp. Bratislava, April 1977, Geologicky ustav Dionyza Stura, Bratislava.* 35–48.

Mock, R. (1978). Novew poznatky o juznych castiach Zapadnych Karpat. In: Vozar, J. (ed.) *Paleogeographical evolution of the West Carpathians, Proc. Symp. Bratislava, April 1977, Geologicky ustav Dionyza Stura, Bratislava.* 321–342.

Moeri, E. (1977). Oberkretazische Schelfsedimente in den Zentralpyrenäen zwischen Rio Segre und Llobregat. *Eclog. geol. Helv.* **70** (1), 193–235.

Mohr, P.A. (1961). *The geology of Ethiopia.* Univ. Coll. Addis Ababa Press, 268pp.

Molenaar, C.M. (1983). Major depositional cycles and regional correlations of Upper Cretaceous rocks, Southern Colorado Plateau and adjacent areas. In: Reynolds, M.W. and Dolly, E.D. (eds.) *Mesozoic Paleogeography of the West-Central United States. Proc. 2nd Rocky Mt. Paleogeogr. Symp., Denver, Colorado. Rocky Mt. Sect. Soc. econ. Paleontol. Mineral.* 201–224.

Mombru, C.A., Uliana, M.A. and Bercowski, F. (1978). Estratigrafía y sedimentología de las acumulaciones biocarbonáticas del Cretácico Inferior sur-mendocino. In: *Acta. 7th Congr. Geol. Argentina, Neuquén, April 1978, I, Sect. III, Geología histórica y regional.* 685–700.

Monciardini, C. (1978). Biozones de Foraminifères et facies du Turonien dans le nord du Bassin de Paris. *Bull. Bur. Rech. géol. minières. sér. 2, sect. I,* **3**, 207–224.

Mongin, D. (1965). Répartition biogéographique des Naïades (La-mellibranches limniques) pendant le Wealdien. In: *Proc. Colloque sur le Crétacé Inférieur, Lyon, Sept. 1963, Mém. Bur. Rech. géol. minières.* **34**, 409–418.

Mongin, D. and Peybernes, B. (1981). L'Albien du chaînon de Fontfroide (zone Prépyrénéenne, sud de la France): observations paléoécologiques sur le gisement de Fontcouverte par l'étude des Mollusques. *Palaeogeogr. Palaeoclimatol. Palaeoecol.* **32** (3-4), 227–246.

Mongin, D., Peybernes, B., Souquet, P. and Thomel, G. (1983). Le Gisement Vraconnien (Albien Supérieur) de la selva de Bonansa (Pyrénées Espagnoles): intérêt stratigraphique, paléoécologique et paléobiogéographique. *Palaeogeogr. Palaeoclimatol. Palaeoecol.* **41** (1), 45–63.

Monleau, C. and Philip, J. (1972). Reconstitution paléogéographique des formations calcaires à rudistes du turonien supérieur de la basse vallée du Rhône à partir d'une étude de microfacies. *Rev. Micropaléontol.* **15** (1), 45–56.

Montadert, L. and Winnock, E. (1971). L'Histoire structurale du Golfe de Gascogne. In: *Proc. Symp. Histoire structurale du Golfe de Gascogne, Rueil-Malmaison, Déc. 1970, Publ. Inst. Fr. Pét., collection colloq. et séminaires.* **22** (2) Editions Technip., Paris, VI-16, 1–18.

Montenat, C. and Bassoullet, J-P. (1983). Le Jurassique et le Néocomien d'Afghanistan central – stratigraphie, évolution paléogéographique. *Eclog. geol. Helv.* **76** (1), 197–241.

Montenat, C. and Vachard, D. (1980). Le Trias des Montagnes Centrales et autres régions d'Afghanistan. *Eclog. geol. Helv.* **73** (3), 697–725.

Moore, C.H. and Martin, K.G. (1966). Comparison of quartz and carbonate shallow marine sandstones, Fredericksburg Cretaceous, central Texas. *Bull. Am. Assoc. Petrol. Geol.* **50** (5), 981–1000.

Moreau, P. (1978). Le rôle du plateau central dans la paléogéographie nord-aquitaine au Cénomanien. In: Philip, J. (ed.) *Colloque sur le Cénomanien: France-Europe occidentale. Paris, Sept. 1976, Géol. Méditerr., Univ. Provence.* **5** (1), 125–136.

Moreno, J.A. (1970). Estratigrafía y paleogeografía del Cretácico Superior en la cuenca del noroeste argentino, con especial mención de los subgrupos Balbuena y Santa Bárbara. *Rev. Assoc. geol. Argentina.* **25** (1), 9–44.

Mork, A., Knarud, R. and Worsley, D. (1982). Depositional and diagenetic environments of the Triassic and Lower Jurassic of Svalbard. In: Embry, A.F. and Balkwill, H.R. (eds.) *Arctic geology and geophysics, Proc. 3rd Int. Symp. Arctic geol., Calgary, June 1981, Mem. Can. Soc. Pet. geol.* **8**, 371–398.

Moroz, V.F. and Romanov, L.F. (1973). New data on Tithonian marine sediments of the Dniester-Prut interfluve. *Dokl. Acad. Sci. USSR.* **204**, 75–78.

Morris, K.A. (1979). A model for the deposition of bituminous shales in the Lower Toarcian. In: *La sédimentation Jurassique W. européen. Proc. Symp., Paris, May 1977, Assoc. Sédimentol. Fr. Spéc. Publ.* **1**, 397–406.

Morris, R.C. and Aleman, A.R. (1975). Sedimentation and tectonics of Middle Cretaceous Copa Sombrero Formation in northwest Peru. *Bol. Soc. geol. Peru.* **48** (1), 49–64.

Moullade, M. (1973). The Ligurian Sea and the adjacent areas. In: Nairn, A.E.M., Kanes, W.H. and Stehli, F.G. (eds.) *Ocean basins and margins,* **4B**, *The West Mediterranean.* Plenum Press, New York, 67–148.

Mouterde, R., Rocha, R.B., Ruget, C. and Tintant, H. (1979). Facies, biostratigraphie et paléogéographie du Jurassique Portugais. In: *Sobre o 1st coloquio de estratigrafia e paleogeografia, de Meso-Cenozoico Portuguese, Lisboa April 1978. Cienc. Terra (Fac. Cienc. Tec., Univ. Nova, Lisbon).* **5**, 29–52.

Mouterde, R. and Ruget, C. (1975). Esquisse de la paléogéographie du Jurassique Inférieur et Moyen au Portugal. *Bull. Soc. géol. Fr. Sér. 7.* **17** (5), 779–786.

Mouterde, R. and Tintant, H. (1968). Le détroit morvano-vosgien a-t-il existe au Lias? In: Rat, P. (ed.) Sud-est du bassin de Paris et ses abords à l'ère secondaire. *Bull. Soc. géol. Fr. Sér. 7.* **10** (3), 293–302.

M'Rabet, A. (1981). Differentiation of environments of dolomite formation, Lower Cretaceous of central Tunisia. *Sedimentol.* **28** (3), 331–352.

Mudge, D.C. (1979). The Cretaceous/Tertiary boundary in the northern North Sea. In: Christensen, W.K. and Birkelund, T. (eds.) *Proc. Cretaceous/Tertiary boundary events symp., Copenhagen Sept. 1979.* **2**, Univ. Copenhagen, 95–102.

Mukherjee, M.K. (1983). Petroleum prospects of Cretaceous sediments of the Cambay Basin, Gujarat, India. *J. Petrol. Geol.* **5** (3), 275–286.

Muller, D. (1973). Perm und Trias im Valle del Bazlan. *Neues Jahrb. Geol. Paläontol. Abhandlungen.* **142** (1), 30–43.

Mullins, H.T. and Lynts, G.W. (1977). Origin of the northwestern Bahama platform: review and reinterpretation. *Bull. geol. Soc. Am.* **88** (10), 1447–1461.

Murat, R.C. (1972). Stratigraphy and palaeogeography of the Cretaceous and Lower Tertiary in southern Nigeria. In: Dessauvagie, T.F.J. and Whiteman, A.J. (eds.) *Proc. Conf. African Geology. Univ. Ibadan, December 1970, Dept. Geol. Univ. Ibadan.* 251–266.

Murris, R.J. (1980). Middle East: stratigraphic evolution and oil habitat. *Bull. Am. Assoc. Petrol. Geol.* **64** (5), 597–618.

Murris, R.J. (1981). Middle East: stratigraphic evolution and oil habitat. *Geol. Mijnb.* **60** (4), 467–486. [N.B. Slightly extended and amended version of Murris (1980)].

Musstow, R. (1968). Stratigraphie und Paläogeographie der Oberkreide und des Albs. *Beih. Geol.* **61**, 71pp.

Mutterlose, J., Schmid, F. and Sporeth, C. (1983). Zur Paläobiogeographie von Belemniten der Unter-Kreide in N.W. Europa. In: *2nd Symp. Kreide, Munich, 1982. Zitteliana.* **10**, 293–308.

Myhr, D.W. and Young, F.G. (1975). Lower Cretaceous (Neocomian) sandstone sequence of Mackenzie Delta and Richardson Mountains area. *Pap. geol. Surv. Can.* **75-1C**, 247–266.

Nachev, I. (1970). Palaeogeography of the Kraishte Region (S.W. Bulgaria) in the Tithonian. *Rev. Bulg. geol. Soc.* **31** (1), 51–62.

Nachev, I.K. (1973). Paleogeography of Bulgaria in the Jurassic (in Russian, English Abstract). *Bull. geol. Inst. Sofia, ser. Stratigr. Lithol.* **22**, 161–194.

Nachev, I.K. (1975). Jurassic paleogeography of the Balkan area (in Russian, English abstract, bilingual captions). *Palaeontol. Stratigr. Lithol.* **11**, 62–76.

Naidin, D.P. (1981). The Russian Platform and the Crimea. In: Reyment, R.A. and Bengtson, P. (eds.) *Aspects of Mid-Cretaceous Regional Geology.* Academic Press, London, 29–68.

Nalivkin, D.V. (1974). Paleogeography of Upper Jurassic and Lower Cretaceous reef massifs in the northern central Tethys. In: *Contributi stratigrafici e paleogeografici sul Mesozoico della Tetide. Mem. Riv. Ital.*

Paleontol. Stratigr. **14**, 307–323.

Nami, M. and Leeder, M.R. (1978). Changing channel morphology and magnitude in the Scalby Formation (Middle Jurassic) of Yorkshire, England. In: Miall, A.D. (ed.) *Fluvial Sedimentology. Mem. Can. Assoc. Petrol. Geol.* **5**, 431–440.

Naylor, D. and Mounteney, S.N. (1975). *Geology of the North West European continental shelf,* 1. Graham Trotman Dudley, 162pp.

Naylor, F.M. (1968). Permian and Triassic formations. In: Sylvester-Bradley, P.C. and Ford, T.D. (eds.) *The geology of the East Midlands.* Leicester University Press, 149–173.

Nelson, E.P., Dalziel, I.W.D. and Milnes, A.G. (1980). Structural geology of the Cordillera Darwin – collisional-style orogenesis in the southernmost Chilean Andes. *Eclog. geol. Helv.* **73** (3), 727–751.

Nelson, H.W. and Glaister, R.P. (1978). Subsurface environmental facies and reservoir relationships of the McMurray oil sands, northeastern Alberta. *Bull. Can. Pet. Geol.* **26** (2), 177–207.

Newkirk, T.F. (1971). Possible future petroleum potential of Jurassic, Western Gulf Basin. In: Cram, I.H. (ed.) *Future petroleum provinces of the United States – their geology and potential. Mem. Am. Ass. Petrol. Geol.* **15** (2), 927–953.

Nichols, K.M. and Silberling, N.J. (1977). Stratigraphy and depositional history of the Star Peak Group (Triassic), northwestern Nevada. *Spec. Pap. geol. Soc. Am.* **178**, 73pp.

Niemczycka, T. (1964). Osady malmu nad górnym Bugiem. *Kwart. geol.* **8** (2), 262–281.

Niemczycka, T. (1976). Jura gorna na obszarze wschodniej Polski (Między Wista a Bugiem). *Pr. Inst. geol. Warszawa.* **77**, 99pp.

Niemczyzka, T. (1978). Litofacje jury środkowej i dolnego oksfordu obszaru radomsko-lubelskiego. *Acta. geol. Pol.* **28** (4), 485–500.

Nikolov, T. (1969). Le Crétacé Inférieur en Bulgarie. *Bull. Soc. géol. Fr. Sér. 7.* **11** (1), 56–68.

Nikolov, T. (1971). Über die Lithofazies der Barreme-Ablagerungen in Bulgarien. *Neues Jahrb. Geol. Paläontol. Abhandlungen.* **139** (2), 163–168.

Nikolov, T.G. (1979). Le cadre paléobiogéographique du Crétacé Inférieur Bulgare. In: Wiedmann, J. (ed.) *Aspekte der Kreide Europas, Proc. 1st Symp. German Cretaceous, Munster, April 1978, Int. Union geol. Sci., Ser. A.* **6**, E. Schweizerbart'sche Verlagsbuchhandlung, Stuttgart, 425–434.

Nilsen, T.H. (1978). Late Cretaceous geology of California and the problem of the proto-San Andreas fault. In: Howell, D.G. and McDougall, K.A. (eds.) *Mesozoic paleogeography of the western United States, Proc. 2nd Symp. Pacific Coast paleogeography, April 1978, Pacific Sect. Soc econ. Paleontol. Mineral.* 559–569.

Nilsen, T.H. and Abbott, P.L. (1981). Paleogeography and sedimentology of Upper Cretaceous turbidites, San Diego, California. *Bull. Am. Assoc. Petrol. Geol.* **65** (7), 1256–1284.

Noldeke, W. (1967). Überblick über die Eisenerzführung in Jura und Kreide des Nordteils der DDR. (Vergleich mit den Angrenzenden Gebieten). *Ber. Deutsch. Ges. geol. Wiss., A., Geol. Paläeontol.* **12** (3–4), 315–327.

Norling, E. (1981). Upper Jurassic and Lower Cretaceous geology of Sweden. *Forh. geol. Foren.* **103** (2), 253–269.

Nwachukwu, S.O. (1975). Sedimentology of Turonian sandstone bodies and paleogeographic conclusions, Aba Omega area, Eastern Nigeria. *Geol. Rdsch.* **64** (2), 330–339.

Obata, I. and Futakami, M. (1977). The Cretaceous sequence of the

Manji Dome, Hokkaido. In: Kanmera, K. (ed.) *Mid-Cretaceous events. Proc. 2nd Int. Conf. Hokkaido, August-Sept. 1976, Spec. Pap. palaeontol. Soc. Japan* **21**, 23–30.

Oberhauser, R. (ed.) (1980). *Der geologische Aufbau Österreichs.* Springer-Verlag, Wien, 701pp.

Ogunyomi, O. and Hills, L.V. (1977). Depositional environments, Foremost Formation (Late Cretaceous), Milk River area, southern Alberta. *Bull. Can. Pet. Geol.* **25** (5), 929–968.

Ojakangas, R.W. (1968). Cretaceous sedimentation, Sacramento Valley, California. *Bull. geol. Soc. Am.* **79** (8), 973–1008.

Ojeda, H.A. (1982). Structural framework, stratigraphy, and evolution of Brazilian marginal basins. *Bull. Am. Assoc. Petrol. Geol.* **66** (6), 732–749.

Ojeda, H.A. and Fugita, A.M. (1974). Bacia Sergipe/Alagoas: geologia regional e perspectivas petroliferas. In: Sanguinetti, Y.T. (ed.) *Proc. 18th Congr. Brazil Geol., Porto Alegre, R.S. October 1974,* 1, *Simposio de recursos energeticos de origem mineral.* 136–158.

Oliver, W.B. (1971). Depositional systems in the Woodbine Formation (Upper Cretaceous) Northeast Texas. *Rep. Invest. Bur. econ. Geol. Univ. Texas.* **73**, 28pp.

Olson, W.S. (1974). Structural history and oil potential of offshore area from Cape Hatteras to Bahamas. *Bull. Am. Assoc. Petrol. Geol.* **58** (6, II), 1191–1200.

Olson, W.S. and Leyden, R.J. (1973). North Atlantic rifting in relation to Permian-Triassic salt deposition. In: Logan, A. and Hills, L.V. (eds.) *The Permian and Triassic systems and their mutual boundary, Proc. Symp. Univ. Calgary, Aug. 1971. Mem. Can. soc. Petrol. Geol.* **2**, 720–732.

Olsson, R.K. (1963). Latest Cretaceous and earliest Tertiary stratigraphy of New Jersey coastal plain. *Bull. Am. Assoc. Petrol. Geol.* **47** (4), 643–665.

Ortlam, D. (1974). Inhalt und Bedeutung fossiler Bodenkomplexe in Perm und Trias von Mitteleuropa. *Geol. Rdsch.* **63** (3), 850–884.

O'Sullivan, R.B. (1970). The upper part of the Upper Triassic Chinle Formation and related rocks, southeastern Utah and adjacent areas. *Prof. pap. U.S. geol. Surv.* **644-E**, 22pp.

Owen, H.G. (1973). Ammonite faunal provinces in the Middle and Upper Albian and their palaeogeographical significance. In: Casey, R. and Rawson, P.F. (eds.) *The Boreal Lower Cretaceous, Geol. J. Spec. Issue.* **5**, Seel House Press, Liverpool, 145–154.

Owen, H.G. (1975). The stratigraphy of the Gault and Upper Greensand of the Weald. *Proc. Geol. Assoc. Lond.* **86** (4), 475–498.

Owen, H.G. (1979). Ammonite Zonal stratigraphy in the Albian of North Germany and its setting in the Hoplitinid faunal province. In: Wiedmann, J. (ed.) *Aspekte der Kreide Europas, Proc. 1st Symp. German Cretaceous, Münster, April 1978, Int. Union geol. Sic. Ser. A.* **6**, E. Schweizerbart'sche Verlagsbuchhandlung, Stuttgart, 563–588.

Ozimic, S. (1982). Depositional environment of the oil shale-bearing Cretaceous Toolebuc Formation and its equivalents, Eromanga Basin, Australia. In: Gary, J.H. (ed.) *15th Oil Shale Symp. Proc., Golden, Colorado April 1982, Colorado School of Mines.* 137–148.

Padula, E. and Mingram, A. (1969). Subsurface Mesozoic red beds of the Choco-Mesapotamian Region, Argentina and their relatives in Uruguay and Brazil. In: *Proc. Ist. Int. Symp. Gondwana Strat., Buenos Aires, 1967, UNESCO.* 1053–1071.

Paine, W.R. and Meyerhoff, A.A. (1970). Gulf of Mexico Basin: interactions among tectonics, sedimentation, and hydrocarbon accumulation. *Trans. Gulf Coast Assoc. geol. Soc.* **20**, 5–44.

Palmer, T.J. (1979). The Hampden Marly and White Limestone Formations: Florida-type carbonate lagoons in the Jurassic of central England. *Palaeontol.* **22** (1), 189–228.

Palmer, T.J. and Jenkyns, H.C. (1975). A carbonate island barrier from the Great Oolite (Middle Jurassic) of central England. *Sedimentol.* **22** (1), 125–135.

Pannekoek, A.J. (ed.) (1956). *Geological history of the Netherlands: explanation to the general geological map of the Netherlands on the scale of 1:200,000.* Govt. Printing and Publishing Office, The Hague, 147pp.

Pareek, H.S. (1979). The Permian-Mesozoic-Palaeogene palaeogeography of the Rajasthan. In: Laskar, B. and Raja Rao, C.S. (eds.) *Proc. 4th Int. Gondwana Symp., Calcutta, 1977.* **1**, Hindustan Publ., Delhi, 23–36.

Pareek, H.S. (1981). Stratigraphy of western Rajasthan. *J. geol. Soc. India.* **22**, 517–527.

Paris, J-P. (1981). Géologie de la Nouvelle-Calédonie. *Mém. Bur. Rech. géol. minières.* **113**, 278pp.

Paris, J-P. and Bradshaw, J.D. (1977). Paléogéographie et géotéctonique de la Nouvelle-Calédonie et de la Nouvelle Zélande au Trias et au Jurassique. *Bull. Bur. Rech. géol. minières., sér. 2, sect. IV,* **1**, 69–78.

Paris, J-P. and Lille, R. (1977). Nouvelle-Calédonie du Permian au Miocene: données cartographiques, hypothèses géotectoniques. *Bull. Bur. Rech. géol. minières., sér. 2, sect. IV,* **1**, 79–95.

Pascal, A. (1976). Variations séquentielles rapides de sédimentation sur la bordure des plates-formes Urgoniennes dans le Nord de l'Espagne. *Bull. Soc. géol. Fr. Sér. 7.* **18** (4), 1005–1008.

Patil, M. (1976). Lithofacies analysis of the Mesozoics of Kutch. *Bull. Oil nat. Gas Comm. Delhi.* **13** (1–2), 33–43.

Patrulius, D. (1969). *Geologia Masivului Becegi: si a culoarului dimbovicioara.* Editura Academiei Republicii Socialiste Romania, Bucuresti, 321pp.

Patrulius, D., Neagu, T., Avram, E. and Pop, G. (1976). The Jurassic-Cretaceous boundary beds in Romania. *Anu. Inst. Geol. Geofiz., Bucharest.* **50**, 71–125.

Pattison, J., Smith, D.B. and Warrington, G. (1973). A review of Late Permian and Early Triassic biostratigraphy in the British Isles. In: Logan, A. and Hills, L.V. (eds.) *The Permian and Triassic systems and their mutual boundary. Proc. Symp. Univ. Calgary, Aug. 1971, Mem. Can. Soc. Petrol. Geol.* **2**, 220–260.

Patzelt, W.J. (1964). Lithologische und paläogeographische Untersuchungen im Unteren Keuper Süddeutschlands. *Erlanger geol. Abhandlungen.* **52**, 30pp.

Paul, J. (1982). Der Untere Buntsandstein des Germanischen Beckens. In: *Proc. Symp. Triassic global geology, Würzburg, Feb. 1982, Geol. Rdsch.* **71** (3), 795–811.

Penn, I.E., Dingwall, R.G. and Knox, R.W.O'B. (1979). The Inferior Oolite (Bajocian) sequence from a borehole in Lyme Bay, Dorset. *Rep. Inst. geol. Sci. Lond.* **79** (3), 27pp.

Penn, I.E. and Evans, C.D.R. (1976). The Middle Jurassic (mainly Bathonian) of Cardigan Bay and its palaeogeographical significance. *Rep. Inst. geol. Sci. Lond.* **76** (6), 8pp.

Peres, F.S. (1968). More information on the Upper Cretaceous lithology of the Moldavian SSR. *Dokl. Acad. Sci. USSR.* 175–177.

Perkins, B.F. (1974). Paleoecology of a rudist reef complex in the Comanche Cretaceous Glen Rose Limestone of Central Texas. In: Perkins, B.F. (ed.) *Aspects of Trinity Division geology: a symposium on the stratigraphy, sedimentary environments and fauna of the Comanche Cretaceous Trinity Division (Aptian and Albian) of Texas and northern Mexico. Geoscience and Man.* **8**, 131–173.

Perrodon, A. (1972). Mécanisme de formation des bassins et dépôts salins. In: Richter-Bernburg, G. (ed.) *Geology of saline deposits, Proc. Symp. Int. Union geol. Sci., Hanover, May 1968, UNESCO, Paris.* 13–18.

Peterson, J.A. (1957). Marine Jurassic of northern Rocky Mountains and Williston Basin. *Bull. Am. Assoc. Petrol. Geol.* **41** (3), 399–440.

Peterson, J.A. (1972). Jurassic system. In: *Geologic Atlas of the Rocky Mountain Region, U.S.A., Rocky Mt. Assoc. Geol.* 177–189.

Petri, S. and Campanha, V.A. (1981). Brazilian continental Cretaceous. In: Mabesoone, J.M., de Brito Neves, B.B. and Sial, A.N. (eds.) *The geology of Brazil. Earth Sci. Rev.* **17** (1), 69–85.

Petters, S.W. (1978). Stratigraphic evolution of the Benue Trough and its implications for the Upper Cretaceous paleogeography of West Africa. *J. Geol.* **86** (3), 311–322.

Petters, S.W. (1979a). Stratigraphic history of the south-central Saharan region. *Bull. geol. Soc. Am.* **90** (8), 753–760.

Petters, S.W. (1979b). Maastrichtian arenaceous foraminifera from north-western Nigeria. *Palaeontol.* **22** (4), 947–963.

Petters, S.W. (1983). Littoral and anoxic facies in the Benue Trough. *Bull. Cent. Rech. Explor. Prod. Elf-Aquitaine.* **7** (1), 361–365.

Petters, S.W. and Ekweozor, C.M. (1982). Origin of mid-Cretaceous black shales in the Benue Trough, Nigeria. *Palaeogeogr. Palaeoclimatol. Palaeoecol.* **40** (4), 311–319.

Peybernès, B. (1978). Dans les Pyrénées la paléogéographie antécénomanienne infirme la théorie d'un coulissement senestre de plusieurs centaines de Kilomètres le long de la faille nord-pyrénéenne des auteurs. *Bull. Soc. géol. Fr. Sér. 7.* **20** (5), 701–709.

Peybernès, B. (1982). Evolution spatio-temporelle des plates-formes carbonatés et des bassins terrigènes dans le Crétacé Inférieur des Pyrénées Franco-Espagnoles. In: Cotillon, P. and Phillip, J. (eds.) *Proc. Mtg. Les facies de plate-forme du Crétacé dans le domaine Mésogéen, 26th Int. geol. Congr., Paris, 1980, Cret. Res.* **3** (1–2), 57–68.

Philip, J. (1964). Facies et variations du Sénonien dans le massif de la Ste-Baume (Var). *Bull. Soc. géol. Fr. Sér. 7.* **6**, suppl., 107–108.

Philip, J. (1972). Paléoécologie des formations à rudistes du Crétacé Supérieur – l'exemple du sud-est de la France. *Palaeogeogr. Palaeoclimatol. Palaeoecol.* **12** (3), 205–222.

Philip, J. (1974). Les formations calcaires à Rudistes du Crétacé Supérieur provencal et rhodanien: stratigraphie et paléogéographie. *Bull. Bur. Rech. géol. minières., Sér.2, Sect.I,* **3**, 107–151.

Philip, J. (1981). Les rudistes du Crétacé Moyen de la Province Méditerranéenne occidentale. Evolution, paléoécologie, paléobiogéographie. *Cret. Res.* **2**, 395–403.

Philip, J. (1982). Paléobiogéographie des Rudistes et géodynamique des marges mésogéennes au Crétacé supérieur. In: *Proc. Symp. Biogéographie et tectonique des plaques, Brest, Mar. 1982. Bull. Soc. géol. Fr. Sér. 7.* **24** (5–6), 995–1006.

Phillip, J. and Allemann, J. (1982). Comparison entres les plates-formes du Crétacé Supérieur de Provence et Sardaigne. In: Cotillon, P. and Phillip, J. (eds.) *Proc. Mtg. Les facies de plate-forme du Crétacé dans le domaine Mésogéen, 26th Int. geol. Congr., Paris, 1980, Cret. Res.* **3** (1–2), 35–46.

Bibliography

Philip, J., Neumann, M. and Porthault, B. *et al.* (1978). Conclusions générales. In: Philip, J. (ed.) *Colloque sur le Cénomanien: France-Europe occidentale. Paris, Sept. 1976, Géol. Méditerr. Univ. Provence.* **5** (1), 205–220.

Pilger, R.J. Jnr. (ed.) (1980). The origin of the Gulf of Mexico and the early opening of the Central north Atlantic Ocean: proceedings of a symposium. *Proc. Symp. Baton Rouge, Louisiana, March 1980. School of Geoscience, Louisiana State Univ., Baton Rouge, Louisiana.* 103pp.

Pinar-Erdem, N. and Ilhan, E. (1977). Outlines of the stratigraphy and tectonics of Turkey, with notes on the geology of Cyprus. In: Nairn, A.E.M., Kanes, W.H. and Stehli, F.G. (eds.) *The ocean basins and margins, 4, The eastern Mediterranean.* Plenum Press, London, 277–318.

Pisa, G., Farabegoli, E. and Ott, E. (1978). Stratigrafia e paleogeografia dei Terreni anisici della conca di Agordo e dell'alta val di Zoldo. (Dolomiti sud-orientali). In: *Contributo all'analisi dei bacini sedimentari, Proc. Mtg. Bologna, June 1978. Mem. Soc. geol. Ital.* **18**, 63–92.

Pisa, G., Marinelli, M. and Viel, G. (1980). Infra Raibl Group: a proposal (Southern Calcareous Alps, Italy). In: Gaetani, M. (ed.) *Contributions to the Triassic stratigraphy Proc. Riccardo Assereto and Giulio Pisa field Symp. on Triassic Stratigraphy of the southern Alps, Bergamo, June 1979. Riv. Ital. Paleontol. Stratigr.* **85** (3–4), 983–1002.

Platel, J-P. (1982). Les formations à rudistes du Turonien moyen et supérieur de la plate-forme nord-Aquitaine. In: *Proc. Colloque sur le Turonien, Paris, Oct. 1981. Mém. Mus. Nat. Hist. nat. Paris, Sér. C.* **49**, 91–96.

Playford, P.E., Cope, R.N. and Cockbain, A.E. (1975). Paleogeographic development of western Australia - Phanerozoic. In: *The geology of western Australia. Mem. geol. Surv. West. Aust.* **2**, 451–499.

Plaziat, J-C. (1981). Late Cretaceous to Late Eocene palaeogeographic evolution of southwest Europe. In: Pomerol, C. (ed.) *Paleogene paleogeography and the geological events at the Eocene/Oligocene boundary. Proc. Symp. Int. geol. Congr., Paris, July 1980. Palaeogeogr. Palaeoclimatol. Palaeoecol.* **35** (2), 263–320.

Plotnikova, L.F., Gevork'yan, V.Kh. and Lipnik, Ye.S. (1969). Albian-Cenomanian boundary in the western Black Sea region. *Dokl. Acad. Sci. USSR.* **183**, 121–123.

Poignant, A-F. (1979). Les Algues rouges Crétacées – relations mer boréale – Téthys. In: Wiedmann, J. (ed.) *Aspekte der Kreide Europas, Proc. 1st Symp. German Cretaceous, Munster, April 1978, Int. Union geol. Sci., Ser. A.* **6**, E. Schweizerbart'sche Verlagsbuchhandlung, Stuttgart, 273–278.

Polak, M. (1978). Jura tatrid Malej Magury, Malej a Wielkiej Fatry (Litologia a paleogeografia). *Pr. geol. Krakow.* **70**, 91-114.

Pollard, J.E. (1981). A comparison between the Triassic trace fossils of Cheshire and south Germany. *Palaeontol.* **24** (3), 555–588.

Polsak, A. (1981). Upper Cretaceous biolithic complexes in a subduction zone: examples from the Inner Dinarides, Yugoslavia. In: Toomey, D.F. (ed.) *European fossil reef models. Spec. Publ. Soc. econ. Paleontol. Mineral.* **30**, 447–472.

Polsak, A., Bauer, V. and Sliskovic, T. (1982). Stratigraphie du Crétacé Supérieur de la plate-forme carbonatée dans les Dinarides Externes. In: Cotillon, P. and Phillip, J. (eds.) *Proc. Mtg. Les facies de plate-forme du Crétacé dans le domaine Mésogéen, 26th Int. geol. Congr., Paris, 1980, Cret. Res.* **3** (1–2), 125–134.

Pomerol, C. (1975). *Stratigraphie et paléogéographie: Ere Mésozoique.* Doin éditeurs, Paris, 383pp.

Ponte, C.F. and Asmus, H.E. (1978). Geological framework of the Brazilian continental margin. *Geol. Rdsch.* **67** (1), 201–235.

Poole, E.G. (1979). The Triassic-Jurassic boundary in Great Britain. *Geol. Mag.* **116** (4), 303–311.

Porter, K.W. and Weimer, R.J. (1982). Diagenetic sequence related to structural history and petroleum accumulation: Spindle Field, Colorado. *Bull. Am. Assoc. Petrol. Geol.* **66** (12), 2543–2560.

Porthault, B. (1978). Paléogéographie et répartition des facies au Cénomanien dans le bassin rhodano-vocontien. In: Philip, J. (ed.) *Colloque sur le Cénomanien: France-Europe occidentale. Paris, Sept. 1976, Géol. Méditerr., Univ. Provence.* **5** (1), 173–182.

Poulton, T.P. (1982). Paleogeographic and tectonic implications of the Lower and Middle Jurassic facies patterns in northern Yukon Territory and adjacent Northwest Territories. In: Embry, A.F. and Balkwill, H.R. (eds.) *Arctic geology and geophysics, Proc. 3rd Int. Symp. Arctic geology, Calgary, June 1981, Mem. Can. Soc. Pet. geol.* **8**, 13–27.

Powell, J.D. (1965). Late Cretaceous platform-basin facies, northern Mexico and adjacent Texas. *Bull. Am. Assoc. Petrol. Geol.* **49** (5), 511–525.

Pozaryska, K. and Brochwicz-Lewinski, W. (1975). The nature and origin of Mesozoic and early Cenozoic marine faunal provinces. *Mitt. Geol. Palaeontol. Inst. Univ. Hamburg.* **44**, 207–216.

Price, I. (1976). Carbonate sedimentology in a pre-Upper Cretaceous continental margin sequence, Othris, Greece. *Bull. Soc. géol. Fr. Sér. 7.* **18** (2), 273–279.

Proctor, C.V. (1980). Distribution of Middle Jurassic facies in the east Shetlands Basin and their control on reservoir capability. In: *The sedimentation of the North Sea reservoir rocks, Proc. Symp. Geilo, May 1980, Nor. Petrol. Soc.* **XV1–22**.

Prokopenko, S.B. (1972). New data on the Jurassic-Cretaceous boundary beds in the west Caspian region. *Dokl. Acad. Sci. USSR.* **200**, 91–94.

Prokopenko, S.B. and Gorbunova, Ye.M. (1977). New data on Volgian-Berriasian lagoonal sediments of the Caspian basin. *Dokl. Acad. Sci. USSR.* **223**, 97–99.

Pryor, W.A. (1960). Cretaceous sedimentation in upper Mississippi embayment. *Bull. Am. Assoc. Petrol. Geol.* **44** (9), 1473–1504.

Pszczolkowski, A. (1978). Geosynclinal sequences of the Cordillera de Guaniguanico in western Cuba; their lithostratigraphy, facies development and paleogeography. *Acta geol. Pol.* **28** (1), 1–96.

Puff, P. and Radzinski, K-H. (1976). Paläogeographie und lagerstättenkundliche Bedeutung des Buntsandsteins. *Z. geol. Wiss. DDR.* **4**, 1483–1503.

Pujatte, V. (1981). Sedimentary succession and paleoenvironments within a fault-controlled basin: the 'Wealden' of the Santander area, northern Spain. *Sediment. Geol.* **28** (4), 293–325.

Purser, B.H. (1979). Dogger Plate-forme bourguignonne. In: *La sédimentation Jurassique W. européen, Proc. Symp. Paris, May 1977, Assoc. Sédimentol. Fr. Spéc. Publ.* **1**, 75–84.

Querol, R. (1969). Petroleum exploration in Spain. In: Hepple, P. (ed.) *The exploration for petroleum in Europe and North Africa. A symposium. Inst. Petrol., Lond.* 49–72.

Raczynska, A. (1979). Stratygrafia i rozwoj litofacjalny mtodszej Kredy Dolnej na Nizu Polskim. *Pr. Inst. geol. Warszawa.* **86**, 78pp.

Rad, U. von (1972). Zur Sedimentologie und Fazies des Allgäuer Flysches. *Geologica Bav.* **66**, 92–147.

Rad, U. von and Arthur, M.A. (1979). Geodynamic, sedimentary and volcanic evolution of the Cape Bojador continental margin (N.W. Africa). In: Talwani, M., Hay, W. and Ryan, W.B.F. (eds.) *Deep drilling results in the Atlantic Ocean: continental margins and paleoenvironment. Proc. Symp. Harriman, New York 1978, Am. Geophys. Union, Maurice Ewing Ser.* **3**, 187–204.

Radelli, L. (1967). Géologie des Andes Colombiennes. *Mem. Trav. Lab. Geol. Fac. Sci. Univ. Grenoble.* **6**, 457pp.

Radig, F. (1963). Die Orbitoiden-Kreide am Nordrande der Mulde von Villarcayo (Prov. Burgos, Spanien) und ihre stratigraphische und paläeogeographische Stellung. *Neues Jahrb. Geol. Paläontol. Abhandlungen.* **117**, 251–264.

Radwanski, S. (19XX) Górnokredowe osady w Sudetach i wptyw tektonicki na ich sedymentacje. *Kwart. geol.* **12** (3), 607–619.

Radwanski, S. (1966). Facje osadowe i charakterystyka faunistyczna Gornej Kredy Środkowych Sudetau. *Ann. Soc. geol. Pol.* **36** (2), 99–119.

Radzinski, K-H. (1967). Gliederung und Paläogeographie des Unteren und mittleren Buntsandsteins im südöstlichen Harzvorland. *Geologie.* **16** (6), 637–659.

Rahmani, R.A. and Lerbekmo, J.F. (1975). Heavy mineral analysis of Upper Cretaceous and Paleocene sandstones in Alberta and adjacent areas of Saskatchewan. In: Caldwell, W.G.E. (ed.) *The Cretaceous System in the Western Interior of North America. Proc. Int. Symp. Univ. Saskatchewan, Saskatoon, May 1973, Spec. Pap. geol. Assoc. Can.* **13**, 607–632.

Ramanathan, R. and Das, K. (1980). Palaeogeography of the Cauvery Basin. In: Venkalacha, B.S. (ed.) *Coastal sedimentaries of India south of 18° N latitude, Proc. Mtg. March 1976. Bull. Oil nat. Gas Comm., Delhi.* **17** (1), 240–246.

Ramos, V.A. (1978). Los arrecifes de la Formación Cotdiano (Jurasico Superior) en la Cordillera Patagonica y su significado paleoclimatico. *Ameghiniana.* **15** (1–2), 97–109.

Rand, H.M. and Mabesoone, J.M. (1982). Northeastern Brazil and the final separation of South America and Africa. *Palaeogeogr. Palaeoclimatol. Palaeoecol.* **38** (3–4), 163–183.

Ranke, U., Rad, U. von and Wissmann, G. (1982). Stratigraphy, facies and tectonic development of the on- and offshore Aaiun - Tarfaya Basin: a review. In: Rad, U. von, Hinz, K., Sarnthein, M. and Seibold, E. (eds.) *Geology of the northwest African continental margin.* Springer-Verlag, Berlin, 86–105.

Rasplus, C. (1979). Données nouvelles sur le Sénonien de Touraine. *Bull. Bur. Rech. géol. minières. Sér. 2, Sect. 1,* **3**, 197–204.

Rat, P. (1964). Problèmes du Crétacé Inférieur dans les Pyrénées et le Nord de l'Espagne. *Geol. Rdsch.* **53** (1), 205–220.

Rat, P. (1965). Rapport sur les formations non marines du Crétacé Inférieur francais. In: *Proc. Colloque sur le Crétacé Inférieur, Lyon, Sept. 1963, Mém. Bur. Rech. géol. minières.* **34**, 333–348.

Rat, P. (1968). Sur le comportement du futur seuil de Bourgogne au Crétacé. In: Rat, P. (ed.) *Sud-est du bassin de Paris et ses abords à l'ère secondaire. Bull. Soc. géol. Fr. Sér. 7.* **10** (3), 393–402.

Rat, P. and Pascal, A. (1982). Les plates-formes carbonatées à rudistes (Dites urgoniennes) du Crétacé Inférieur et leur environnement. In: Cotillon, P. and Phillip, J. (eds.) *Proc. Mtg. Les facies de plate-forme du Crétacé dans le domaine Mésogéen, 26th Int. geol. Congr., Paris, 1980, Cret. Res.* **3** (1–2), 155–166.

Rau, A. and Tongiorgi, M. (1980). Alcune ipotesi sulla storia Pre-Giurassica del futuro margine continentale Nord-Appenninico. In: *L'evoluzione tettonico-sedimentaria dell'Appennino settentrionale, 70th Cong. Soc. geol. Ital., Siena, Oct. 1980. Mem. Soc. geol. Ital.* **21**, 23–30.

Rautman, C.A. (1978). Sedimentology of Late Jurassic barrier- island complex - lower Sundance Formation of Black Hills. *Bull. Am. Assoc. Petrol. Geol.* **62** (11), 2275–2289.

Rawson, P.F. (1973). Lower Cretaceous (Ryazanian-Barremian) marine connections and cephalopod migrations between the Tethyan and Boreal Realms. In: Casey, R. and Rawson, P.F. (eds.) *The Boreal Lower Cretaceous, Geol. J. Spec. Issue* **5**, Seel House Press, Liverpool, 131–144.

Rawson, P.F. (1980). United Kingdom: the Lower Cretaceous. In: Lorenz, C. (ed.) *Geology of the European countries: Austria, Federal Democractic Republic of Germany, Ireland, The Netherlands, Switzerland, United Kingdom. 26th Int. geol. Congr. Paris.* 408–409.

Rayer, F.G. (1981). Exploration prospects and future petroleum potential of the Canadian Arctic Islands. *J. Petrol. Geol.* **3** (4), 367–412.

Reeside, J.B. Jnr. (1957). Paleoecology of the Cretaceous Seas of the Western Interior of the United States. In: Hedgpeth, J.W. and Ladd, H. (eds.) *Treatise on marine ecology and paleoecology, 2, Paleoecology. Mem. geol. Soc. Am.* **67** (2), 505–542.

Reid, R.E.H. (1976). Late Cretaceous climate trends, faunas, and hydrography in Britain and Ireland. *Geol. Mag.* **113** (2), 115–128.

Reif, D.M. and Slatt, R.M. (1979). Red bed members of the Lower Triassic Moenkopi Formation, southern Nevada: sedimentology and paleogeography of a muddy tidal flat deposit. *J. sediment. Petrol.* **49** (3), 869–890.

Reif, W-E. (1971). Zur Genese des Muschelkalk – Keuper – Grenzbonebeds in Südwestdeutschland. *Neues Jahrb. Geol. Paläontol. Abhandlungen.* **139** (3), 369–404.

Reiskind, J. (1975). Marine concretionary faunas of the upper Bearpaw Shale (Maestrichtian) in eastern Montana and southwestern Saskatchewan. In: Caldwell, W.G.E. (ed.) *The Cretaceous System in the Western Interior of North America. Proc. Int. Symp. Univ. Saskatchewan, Saskatoon, May 1973, Spec. Pap. geol. Assoc. Can.* **13**, 235–252.

Reitner, J. and Wiedmann, J. (1982). Die Deltaphasen der Unterkreide am Basko-Kantabrischen Kontinentalrand, Nordspanien. In: Wiedmann, J., Kullmann, J. and Schonenberg, R. (eds.) *Subsidenz-Entwicklung im Kantabrischen Variszikum und an passiven Kontinentalrandern der Kreide, 2, Kreide, Neues Jahrb. Geol. Paläont. Abhandlungen.* **165** (1), 60–76.

Remane, J. (1966). Note préliminaire sur la paléogéographie du Tithonique des chaînes subalpines. *Bull. Soc. géol. Fr. Sér. 7.* **8** (3), 448–453.

Repenning, C.A. and Page, H.G. (1956). Late Cretaceous stratigraphy of Black Mesa, Navajo and Hopi Indian reservations, Arizona. *Bull. Am. Assoc. Petrol. Geol.* **40** (2), 255–294.

Rey, J. (1972). Recherches géologiques sur le Crétacé Inférieur de l'Estramadura (Portugal). *Mem. Serv. geol. Portugal.* **21**, 477pp.

Rey, J. (1979). Les formations bioconstruites du Crétacé Inférieur d'Estramadura (Portugal). In: *L'Urgonien des Pays Méditerranéens. Géobios Spéc. Mém.* **3**, 89–102.

Rey, J. (1979). Le Crétacé Inférieur de la marge Atlantique Portugaise; biostratigraphie, organisation séquentielle, évolution paléogéographique. In: *Sobre O 1st coloquio de estratigrafria e paleogeografia de Meso-Cenozoico Portuguese, Lisboa, April 1978. Cienc. Terra (Fac. cienc. Tec., Univ. Nova, Lisboa).* **5**, 97–120.

Rey, J., Bilotte, M. and Peybernes, B. (1977). Analyse biostratigraphique et paléontologique de l'Albien marin d'Estremadura (Portugal). *Geobios.* **10** (3), 369–393.

Reyes, F.C. and Salfity, J.A. (1973). Consideraciones sobre la estratigrafía del Cretácico (Subgrupo Pirgua) del noroeste de Argentina. In: *Proc. 5th Congr. Geológico Argentino, Ciudad de Villa Carlos Paz (Cordoba), October 1972, 3, Sect. 5, Geología histórica y estratigrafía, Librart, Buenos Aires.* 355–385.

Reyment, R.A. (1974). Application des méthodes paléobiologiques à la théorie de la dérive des continents, illustrée par l'Atlantique Sud. *Rev. Géogr. phys. Géol. dyn.* **16** (1), 61–70.

Reyment, R.A. (1980a). Beobachtungen über die Ostrakoden der Coniac Transgression im Saharagebiet. *Neues Jahrb. Geol. Paläontol. Monatshefte.* **1980** (4), 252–256.

Reyment, R.A. (1980b). Biography of the Saharan Cretaceous and Palaeocene epicontinental Transgression. *Cret. Res.* **1** (4), 299–327.

Reyment, R.A. (1981). West Africa. In: Reyment, R.A. and Bengtson, P. (eds.) *Aspects of Mid-Cretaceous Regional Geology.* Academic Press, London, 133–160.

Reyment, R.A. Bengtson, P. and Tait, E.A. (1976). Cretaceous transgressions in Nigeria and Sergipe-Alagoas (Brazil). In: Almeida, F.F.M. (ed.) *Simposio internacional sobre as margens continentais de tipo Atlantico, Proc. Symp. Sao Paulo, October 1975, An. Acad. bras. Cienc.* **48**, Suppl., 253–264.

Reyment, R.A. and Morner, N.A. (1977). Cretaceous transgressions and regressions exemplified by the South Atlantic. In: Kanmera, K. (ed.) *Mid-Cretaceous events. Proc. 2nd Int. Conf., Hokkaido, Aug-Sept 1976. Spec. Pap. palaeontol. Soc. Japan.* **21**, 247–261.

Reyre, D. (1966). Particularités géologiques des bassins côtiers de l'Ouest Africain. In: Reyre, D. (ed.) *Sedimentary Basins of the African Coasts, 1, Atlantic Coast. Proc. Symp. New Delhi, 1964, Assoc. Serv. geol. Afr., Paris.* 250–301.

Reyre, D., Belmonte, Y., Derumaux, F. and Wenger, R. (1966). Evolution géologique du bassin gabonnais. In: Reyre, D. (ed.) *Sedimentary basins of the African Coast, 1, Atlantic Coast Proc. Symp. New Delhi, 1964, Assoc. Serv. geol. Afr., Paris.* 171–190.

Riba, O. and Rios, J.M. (1962). Observations sur la structure du secteur sud-ouest de la Chaîne ibérique (Espagne). In: Durand-Delga, M. (ed.) *Livre à la Mémoire du Professeur Paul Fallot: L'évolution paléogéographique et structurale des domaines méditerranéens et Alpins d'Europe. Mém. Soc. géol. Fr. Hors Sér.* **1**, 275–290.

Ricateau, R. and Villemin, J. (1973). Evolution au Crétacé Supérieur de la pente separant le domaine de plate-forme du sillon sous-Pyrénéen en Aquitaine méridionale. *Bull. Soc. géol. Fr. Sér.* 7. **15** (1), 30–39.

Riccardi, A.C. (1976). Paleontología y edad de la Formación Springhill. In: *Proc. 1st Congr. Geol. Chileno, Santiago, Aug. 1976, 1, Sect. C, Palaeontol.* C41–56.

Rice, D.D. (1980). Coastal and deltaic sedimentation of Upper Cretaceous Eagle sandstone: relation to shallow gas accumulations, north-central Montana. *Bull. Am. Assoc. Petrol. Geol.* **64** (3), 316–338.

Rice, D.D. and Shurr, G.W. (1980). Shallow, low-permeability reservoirs of northern Great Plains – assessment of their natural gas resources. *Bull. Am. Assoc. Petrol. Geol.* **64** (7), 969–987.

Rice, D.D. and Shurr, G.W. (1983). Patterns of sedimentation and paleogeography across the Western Interior Seaway during time of deposition of Upper Cretaceous Eagle sandstone and equivalent rocks, northern Great Plains. In: Reynolds, M.W. and Dolly, E.D. (eds.) *Mesozoic Paleogeography of the West-Central United States, Proc. 2nd Rocky Mt. Paleogeogr. Symp., Denver, Colorado. Rocky Mt. Sect., Soc. econ. Paleontol. Mineral.* 337–358.

Rich, E.I. (1958). Stratigraphic relation of latest Cretaceous rocks in parts of Powder River, Wind River and Big Horn Basins, Wyoming. *Bull. Am. Assoc. Petrol. Geol.* **42** (10), 2424–2443.

Richards, H.G. (1974). Tectonic evolution of Alaska. *Bull. Am. Assoc. Petrol. Geol.* **58** (1), 79–105.

Riche, P. and Prestat, B. (1980). Paléogéographie du Crétacé Moyen du proche et Moyen-Orient et sa signification pétrolière. In: *Proc. 10th World Petrol. Congr. Bucharest, 1979, 2, Exploration, supply and demand.* Heyden & Son, London, 57–75.

Richter-Bernburg, G. (1972). Saline deposits in Germany; a review and general introduction to the excursions. In: Richter-Bernburg, G. (ed.) *Geology of saline deposits, Proc. Symp. Int. Union geol. Sci., Hanover, May 1968, UNESCO, Paris.* 275–287.

Ricour, J. (1963). Particularités paléogéographiques des Alpes occidentales francaises aux temps Triasiques. In: Durand-Delga, M. (ed.) *Livre à la Mémoire du Professeur Paul Fallot: L'évolution paléogéographique et structurale des domaines méditerranéens et Alpins d'Europe. Mém. Soc. géol. Fr. Hors Sér. 1.* **2**, 395–406.

Ridd, M.F. (1978). Thailand. In: Moullade, M. and Nairn, A.E.M. (eds.) *The Phanerozoic geology of the world, 2, The Mesozoic, A.* Elsevier, Amsterdam, 145–154.

Rieber, H. and Sorbini, L. (1983). Middle Triassic bituminous shales of Monte San Giorgia (Tessin, Switzerland). *Guide Book to Excursion 11A 1st Int. Congr. Paleocol. Lyon, July 1983.* 40pp.

Riech, V. and Rad, U. von. (1979). Silica diagenesis in the Atlantic Ocean: diagenetic potential and transformations. In: Talwani, M., Hay, W. and Ryan, W.B.F. (eds.) *Deep drilling results in the Atlantic Ocean: continental margins and paleoenvironment. Am. Geophys Union, Washington D.C., Maurice Ewing Ser.* **3**. 315–340.

Riegraf, W. and Keller, T. (1979). Zwei seltene Ammonitenarten im Posidonienschiefer (Untertoarcium) Süddeutschlands. *Neues Jahrb. Geol. Paläontol. Monatshefte.* **1979** (11), 690–700.

Rios, J.M. (1978). The Mediterranean coast of Spain and the Alboran Sea. In: Nairn, A.E.M., Kanes, W.H. and Stehli, F.G. (eds.) *The ocean basins and margins, 4B, The western Mediterranean.* Plenum Press, London, 1–66.

Robaszynski, F. (1979). Comparison between the Middle Cretaceous of Belgium and some French regions. In: Wiedmann, J. (ed.) *Aspekte der Kreide Europas, Proc. 1st Symp. German Cretaceous, Münster, April 1978, Int. Union geol. Sci., Ser. A.* **6**, E. Schweizerbart'sche Verlagsbuchhandlung, Stuttgart, 543–562.

Robaszynski, F. (1981). Répartitions comparées de foraminifères planctoniques du Crétacé Moyen dans la Téthys et la Mer Boréale par rapport à une zonéologie macropaléontologique. *Cret. Res.* **2** (3–4), 409–416.

Roberts, D.G. (1974). Structural development of the British Isles, the continental margin, and the Rockall Plateau. In: Burk, C.A. and Drake, C.L. (eds.) *The geology of continental margins.* Springer-Verlag, New York, 343–360.

Roberts, R.J. (1972). Evolution of the Cordilleran fold belt. *Bull. geol. Soc. Am.* **83** (7), 1989–2004.

Robinson, E. (1967). Submarine slides in White Limestone Group,

Jamaica. *Bull. Am. Assoc. Petrol. Geol.* **51** (4), 569-578.

Robinson, P.L. (1969). The Indian Gondwana Formations – a review. *Proc. 1st Int. Symp. Gondwana Strat. Paleontol., Buenos Aires, 1967, UNESCO.* 201–268.

Robinson, P.L. (1971). A problem of faunal replacement on Permo-Triassic continents. *Palaeontol.* **14** (1), 131–153.

Rocha-Campos, A.C. (1973). Upper Paleozoic and Lower Mesozoic paleogeography and paleoclimatological and tectonic events in South America. In: Logan, A. and Hills, L.V. (eds.) *The Permian and Triassic systems and their mutual boundary, Proc. Symp. Univ. Calgary, Aug. 1971, Mem. Can. Soc. Petrol. Geol.* **2**, 398–424.

Rodríguez Estrella, T. (1977). Síntesis geológica del Prebético de la Provincia de Alicante. 1, Estratigrafía. *Bol. Geol. Min. Inst. Geol. Min. España.* **88** (3), 183–214.

Rogl, F. (1980). Biostratigraphie und Paläogeographie. In: Bachmayer, F. (ed.) *Erdöl und Erdgas in Österreich. Veröff. naturhist. Mus. Wien, N.F.* **19**, 247–251.

Rolle, F., Koch, J-O., Frandsen, N. and Surlyk, F. (1979). Jurassic environments in the Fenno-Scandian border zone. In: *La sédimentation Jurassique W.européen Proc. Symp., Paris May 1977, Assoc. Sédimentol. Fr. Spéc. Publ.* **1**, 15–31.

Rolleri, E.O. and Roque, P.C. (1969). Geología de la Provincia de Mendoza. *Actas Cuartas Jornadas geol. Argentinas, Mendoza, April 1969, Librart, Buenos Aires.* **2**, 1–60.

Rollet, M. (1969). Recherches géologiques dans la Skopska Crna Gora-Karadagh (Macédoine Yougoslave). *Ann. sci. Univ. Besancon, Sér. 3.* **12**, 332pp.

Ronnevik, H., Eggen, S. and Vollset, J. (1983). Exploration of the Norwegian Sea. In: Brooks, J. (ed.) *Petroleum geochemistry and exploration of Europe. Proc. Symp. Br. geol. Congr., Univ. Glasgow, Sept. 1982. Spec. Publ. geol. Soc. Lond.* **12**, 71–94.

Rose, P.R. (1972). Edwards Group, surface and subsurface, Central Texas. *Rep. Invest. Bur. econ. Geol. Univ. Texas.* **74**, 198pp.

Rosenfeld, U. (1978). Beitrag zur Paläogeographie des Mesozoikums in Westfalen. *Neues Jahrb. Geol. Paläontol. Abhandlungen.* **156** (1), 132–155.

Rosenfeld, U. (1981). Das Südende der Anden-Geosynklinale. In: Miller, H. and Rosenfield, U. (eds.) *Beiträge zum VII Geowissenschaftlichen Lateinamerika – Kolloquium, Heidelberg 1980, Zentralbl. Geol. Paläontol. 1981.* **1** (3–4), 333–343.

Rosenfeld, U. and Volkheimer, W. (1980). Turbidite und andere Rhythmite im tieferen Jura des Neuquen-Beckens (Argentinien). *Neues Jahrb. Geol. Paläontol. Abhandlungen.* **159** (3), 379–421.

Rousset, C. (1973). Rôle de la Karstologie dans l'élaboration des reconstitutions paléogéographiques: les cycles Karstiques en Provence. *Rev. Géogr. phys. Géol. dyn.* **15** (3), 273–294.

Rudkin, R.A. (1964). Lower Cretaceous. In: McGrossan, R.G. and Glaister, R.P. (eds.) *Geological history of western Canada. Alberta Soc. Petrol. Geol.* 156–168.

Rumohr, J. (1973). Deltaisch-fluviatile Sedimentation des tiefen Malm (Wiehengebirgsquarzit) am Gehn (Wiehengebirge, Niedersachsen). *Neues Jahrb. Geol. Paläontol. Abhandlungen.* **143** (3), 345–383.

Rusitzka, D. (1967). Paläogeographie der Trias im Nordteil der DDR. *Ber. Deutsch. Ges. geol. Wiss., A., Geol. Palaeontol.* **12** (3–4), 243–257.

Rusitzka, D. and Jubitz, K-B. (1968). Trias. In: Schmidt, K. (ed.) *Grundriss der Geologie der Deutschen Demokratischen Republik, 1, Geologische Entwicklung des Gesamtgebietes.* Akademie Verlag, Berlin. 268–

289 (+ Anl.-Abb. 11–13).

Ruskova, N. (1975). Lithofacies zoning of the Lower Cretaceous in central northern Bulgaria. *Geologica Balc.* **5** (3), 55–63.

Ryder, R.T. and Scholten, R. (1973). Syntectonic conglomerates in southwestern Montana: their nature, origin and tectonic significance. *Bull. Geol. Soc. Am.* **84** (3), 773–796.

Ryer, T.A. (1977). Patterns of Cretaceous shallow-marine sedimentation, Coalville and Rockport areas, Utah. *Bull. Geol. Soc. Am.* **88** (2), 177–188.

Ryer, T.A. and McPhillips, M. (1983). Early Late Cretaceous paleogeography of east-central Utah. In: Reynolds, M.W. and Dolly, E.D. (eds.) *Mesozoic Paleogeography of the West-Central United States. Proc. 2nd Rocky Mt. Paleogeogr. Symp., Denver, Colorado. Rocky Mt. Sect., Soc. econ. Paleontol. Mineral.* 253–272.

Sabins, F.F. (1963). Anatomy of stratigraphic trap, Bisti field, New Mexico. *Bull. Am. Assoc. Petrol. Geol.* **47** (2), 193–228.

Sachs, V.N., Basov, V.A. and Dagis, A.A. (1973). Palaeozoogeography of Boreal-Realm Seas in Jurassic and Neocomian. In: Pitcher, M.G. (ed.) *Arctic Geology. Proc. 2nd Int. Symp. Arctic Geology, February 1971, San Francisco, Mem. Am. Assoc. Petrol. Geol.* **19**, 219–229.

Sachs, V.N. and Strelkov, S.A. (1961). Mesozoic and Cenozoic of the Soviet Arctic. In: Raasch, G.O. (ed.) *Geology of the Arctic, Proc. 1st Int. Symp. Arctic geology, Alberta, 1960.* **1**, Univ. Toronto Press, 48–67.

Saint-Marc, P. (1969). Etude géologique de la région d'Hermel (Liban septentrional). *Bull. Soc. géol. Fr. Sér. 7.* **11** (3), 379–387.

Saint-Marc, P. (1978). Arabian Peninsula. In: Moullade, M. and Nairn, A.E.M. (eds.) *The Phanerozoic geology of the world, 2, The Mesozoic, A.* Elsevier, Amsterdam, 435–462.

Salaj, J. (1978). The geology of the Pelagian Block: the eastern Tunisian Platform. In: Nairn, A.E.M., Kanes, W.H. and Stehli, F.G. (eds.) *The ocean basins and margins, 4B, The western Mediterranean.* Plenum Press, London, 361–416.

Salaj, J. (1980). Microbiostratigraphie du Crétacé et du Paléogène de la Tunisie septentrionale et orientale (Hypostratotypes Tunisiens). *Inst. geol. Dionyz Stur, Bratislava.* 238pp.

Salaj, J. and Bellier, J-P. (1976). Une coupe de référence pour la zonation de l'Albien, du Cénomanien et du Turonien de Tunisie septentrionale. In: *Evénements de la partie moyenne du Crétacé: rapports sur la biostratigraphie des régions clés, Uppsala 1975 – Nice 1976, Ann. Mus. Hist. nat. Nice.* **4**, XXI 1–10.

Salas, G.P. (1968). Petroleum evaluation of North-Central Mexico. *Bull. Am. Assoc. Petrol. Geol.* **52** (4), 665–674.

Salomon, J. (1983). Les phases 'fosse' dans l'histoire du Bassin de Soria (Espagne du Nord) au Jurassique supérieur-Crétacé Inférieur. *Bull. Cent. Rech. Explor. Prod. Elf-Aquitaine.* **7** (1), 399–407.

Salvador, A. and Green, A.R. (1980). Opening of the Caribbean Tethys (origin and development of the Caribbean and the Gulf of Mexico). In: Aubouin, J., Debelmas, J. and Latreille, M. (eds.) *Géologie des chaînes alpines issues de la Téthys, Colloque. C5, 26th Int. geol. Congr., Paris, 1980, Mém. Bur. Rech. géol. minières.* **115**, 224–229.

Salvan, H.M. (1974). Les séries salifères du Trias marocain; caractères généraux et possibilités d'interprétation. *Bull. Soc. géol. Fr. Sér. 7.* **16** (6), 724–731.

Sames, C-W. (1970). Orogene Sedimentation und Paläogeographie während der tieferen Oberkreide in Norditalien. *Beih. geol. Jahrb.* **103**, 115pp.

Bibliography

Sander, N.J. (1970). Structural evolution of the Mediterranean region during the Mesozoic era. In: Alvarez, W. and Gohrbandt, K.H. (eds.) *Geology and History of Sicily, 12th Ann. Field Conf., Petrol. Explor. Soc. Libya*. 43–132.

Sapunov, I.G. (1971). Notes on the geographical differentiation of the Lower Jurassic Ammonite faunas. *Proc. 2nd Int. Colloque du Jurassique, Luxembourg, 1967. Mém. Bur. Rech. géol. minières*. **75**, 263–270.

Sastri, V.V. and Raiverman, V. (1968). On the basin study programme of the Cretaceous-Tertiary sediments of the Cauvery Basin. In: *Cretaceous-Tertiary Formations of South India, Proc. seminar Bangalore, June 1966, Mem. geol. soc. India*. **2**, 143–152.

Sastri, V.V., Raju, A.T.R., Sinha, R.N., Venkatachala, B.S. and Banerji, R.K. (1977). Biostratigraphy and evolution of the Cauvery Basin, India. In: *Oil exploration in the east coast of India: stratigraphic and paleontological aspects. J. geol. Soc. India*. **18** (8), 355–377.

Sato, T. (1961). Etudes biostratigraphiques des ammonites du Jurassique du Japon. *Mém. Soc. géol. Fr. Hors Sér*. **94**, 122pp.

Sato, T. (1975). Marine Jurassic Formations and faunas in South East Asia and New Guinea. In: Kobayashi, T. and Toriyama, R. (eds.) *The Geology and Palaeontology of South East Asia*. **15**, Univ. Tokyo Press, 151–190.

Schafer, K.A. (1973). Facies and paleogeography of the 'Spiriferina-Bank' (Upper Muschelkalk) in Southwest Germany. *Neues Jahrb. Geol. Palaontol. Abhandlungen*. **143** (1), 56–110.

Schlee, J.S. (1981). Seismic stratigraphy of Baltimore Canyon Trough. *Bull. Am. Assoc. Petrol. Geol*. **65** (1), 26–53.

Schloz, W. (1972). Zur Bildungsgeschichte der Oolithenbank (Hettangium) in Baden-Württemberg. *Arb. Inst. Geol. Paläeontol. Univ. Stuttgart, N.F*. **67**, 101–212.

Schmidt, V. (1965). Facies, diagenesis and related reservoir properties in the Gigas Beds (Upper Jurassic), northwestern Germany. In: Pray, L.C. and Murray, R.C. (eds.) *Dolomitisation and limestone diagenesis – a symposium. Spec. Publ. Soc. econ. Paleontol. Mineral*. **13**, 124–168.

Schmidt-Kaler, H. (1962a). Zur Ammonitenfauna und Stratigraphie des Malm Alpha und Beta in der südlichen und Mittleren Frankenalb. *Erlanger geol. Abhandlungen*. **43**, 12pp.

Schmidt-Kaler, H. (1962b). Stratigraphische und tektonische Untersuchungen in Malm des nordöstlichen Ries-Ratimens. *Erlanger geol. Abhandlungen*. **44**, 51pp.

Schmidt-Kaler, H. (1979). *Geologische Karte des Naturparks: Altmühltal Südliche Frankenalb 1:100,000, mit Kurzerläuterungen auf der Rückseite*. Bayerisches Geologisches Landesamt, München.

Schmidt-Thorme, P. (1963). Le bassin de la Molasse d'Allemagne du Sud, avec des considérations particulières sur la Molasse plissée de Bavière. In: Durand-Delga, M. (ed.) *Livre à la Mémoire du Professeur Paul Fallot: L'évolution paléogéographique et structurale des domaines méditerranéens et Alpins d'Europe. Mém. Soc. géol. Fr. Hors Sér.1*, **2**, 431–452.

Schmitt, G.T. (1953). Regional stratigraphic analysis of middle and upper marine Jurassic in northern Rocky Mountains – Great Plains. *Bull. Am. Assoc. Petrol. Geol*. **37** (2), 355–393.

Schopf, T.J.M. (1974). Permo-Triassic extinctions: relation to sea-floor spreading. *J. Geol*. **82** (2), 129–143.

Schott, W. (1949). Zur Paläeogeographie des nordwestdeutschen Malms. In: Bentz, A. (ed.) *Erdöl und Tektonik in Nordwestdeutschland,*

Proc. Symp. Hannover, Sept. 1947, Veröff. Amt für Bodenforsch. 129–135.

Schott, W. (1968). Nordwestdeutsches Wealdenbecken und Ostseebecken (Gedanken zur Paläogeographie des Wealden). *Geol. Jahrb*. **85**, 919–940.

Schott, W., Jaritz, W., Kockel, F., Sames, C.W., Stackelberg, V. von., Stets, J., Stoppel, D., Baldschuhn, R. and Krampe, K-D. (1969). Erläuterungen des Paläogeographischen Atlas der Unterkreide von Nordwestdeutschland: mit einer Übersichtsdarstellung des nördlichen Mitteleuropa., *Bundesanst. für Bodenforsch.*, 315pp.

Schott, W. and Staesche, K. (1967). Zur Stratigraphie der grobklastischen Fazies im Oxfordium des Wiehengebirges. *Geol. Rdsch.* **56** (3), 766–784.

Schroder, B. (1962). Schwermineralführung und Paläogeographie des Doggersandsteins in Nordbayern. *Erlanger geol. Abhandlungen.* **42**, 29pp.

Schroder, B. (1971). Strukturell-fazielle Entwicklung Nord-Bayerns während Trias und Jura. *Neues Jahrb. Geol. Palaontol. Abhandlungen.* **138** (1), 101–118.

Schroder, B. (1982). Entwicklung des Sedimentbeckens und Stratigraphie der klassischen Germanischen Trias. In: *Proc. Symp. Triassic global geology, Würzburg, Feb. 1982, Geol. Rdsch.* **71** (3), 783–794.

Schuchert, C. (1955). *Atlas of paleogeographic maps of North America*. Wiley & Sons. New York, 177pp.

Schuler, G. (1968). Lithofazielle, sedimentologische und paläogeographische Untersuchungen in den Raibler Schichten zwischen Inn und Salsach (Nördliche Kalkalpen). *Erlanger geol. Abhandlungen.* **71**, 60pp.

Schultheis, N.H. and Mountjoy, E.W. (1978). Cadomin conglomerate of western Alberta – a result of Early Cretaceous uplift of the main ranges. *Bull. Can. Pet. Geol.* **26** (3), 297–342.

Schwarz, H-U. (1971). Facies analysis of shallow marine carbonates (Lower Muschelkalk, Middle Triassic). In: Muller, G. (ed.) *Sedimentology of parts of Central Europe, Guidebook to excursions. 8th Int. Sedimentol. Congr., Heidelburg, 1971*. Verlag Waldemar Kramer, Frankfurt, 125–131.

Schweickert, R.A. (1978). Triassic and Jurassic paleogeography of the Sierra Nevada and adjacent regions, California and western Nevada. In: Howell, D.G. and McDougall, K.A. (eds.) *Mesozoic paleogeography of the western United States, Proc. 2nd Symp. Pacific Coast Paleogeogr, April 1978, Pacific Sect. Soc. econ. Paleontol. Mineral.* 361–384.

Sclater, J.G., Boyle, E. and Edmond, J.M. (1979). A quantitative analysis of some factors affecting carbonate sedimentation in the oceans. In: Talwani, M., Hay, W. and Ryan, W.B.F. (eds.) *Deep drilling results in the Atlantic Ocean: continental margins and paleoenvironment. Proc. Symp. Harriman, New York, March 1978. Am. geophys. Union, Maurice Ewing Ser.* **3**, 235–248.

Sclater, J.G., Hellinger, S. and Tapscott, C. (1977). The paleobathymetry of the Atlantic ocean from Jurassic to the present. *J. Geol.* **85** (5), 509–552.

Scott, R.W. (1970). Stratigraphy and sedimentary environments of Lower Cretaceous rocks, southern Western Interior. *Bull. Am. Assoc. Petrol. Geol.* **54** (7), 1225–1244.

Scott, R.W. (1978). Paleobiology of Comanchean (Cretaceous) cardiids (*Cardiinae*), North America. *J. Paleontol.* **52** (4), 881–903.

Scott, R.W. (1979). Depositional model of early Cretaceous coral-algal-rudist reefs, Arizona. *Bull. Am. Assoc. Petrol. Geol.* **63** (7), 1108–

1127.

Sedletskiy, V.I., Derevyagin, V.S. and Omel'chenko, Yu.V. (1976). Upper Jurassic evaporites of Ciscaucasia. *Dokl. Acad. Sci. USSR.* **220**, 96–97.

Seibertz, E. (1977). Litho-, Bio-, Ökostratigraphie, Sedimentologie und Tektonik im Soester Grünsand (Oberes Mittelturon, südliches Münsterland). *Geol. Jahrb.* **A40**, 61–113.

Seidel, G. (1965). Zur geologischen Entwicklungsgeschichte des Thüringer Beckens. *Beih. Geol.* **50**, 115pp.

Seitz, O. (1949). Zur Paläogeographie des Wealden und der Pompeckj'schen Schwelle. In: Bentz, A. (ed.) *Erdöl und Tektonik in Nortwestdeutschland, Proc. Symp. Hannover, Sept. 1947, Veroff. Amt für Bodenforsch.*, 135–143.

Sellwood, B.W. (1972a). Regional environmental changes across a Lower Jurassic stage-boundary in Britain. *Palaeontol.* **15** (1), 125–157.

Sellwood, B.W. (1972b). Tidal-flat sedimentation in the Lower Jurassic of Bornholm, Denmark. *Palaeogeogr. Palaeoclimatol. Palaeoecol.* **11** (1) 93–106.

Sellwood, B.W. and McKerrow, W.S. (1974). Depositional environments in the lower part of the Great Oolite Group of Oxfordshire and north Gloucestershire. *Proc. Geol. Assoc. Lond.* **85** (2), 189–210.

Senkowiczowa, H. (1965). Podziąt i rozwój facjalny osadow retu na obszarze południowej Polski. *Kwart. geol.* **9** (2), 241–260.

Senkowiczowa, H. (1969). Trias (bez utworow retyku). In: *Stratygrafia Mezozoiku obrzeżenia gor Swietokryskich. Pr. Inst. geol. Warszawa.* **56**, 7–48.

Senkowiczowa, H. and Szyperko-Sliwczynska, A. (1975). Stratigraphy and palaeogeography of the Trias. In: Slowanska, B. and Wilczek, M. (eds.) *Proc. 50th Anniv. Symp. geol. Inst., Warsaw, June 1970, Bull. geol. Inst.* **252**, 131–147.

Septfontaine, M. and Lombard, A. (1976). Le Jurassique des Préalpes médianes dans le S.W. du Chablais (Haute-Savoie, France): cadre tectonique et lithostratigraphique. *Eclog. geol. Helv.* **69** (2), 425–460.

Seronie-Vivien, R., Sens, J. and Malmoustier, G. (1965). Contribution à l'étude des formations du Crétacé Inférieur dans le Bassin de Parentis (Aquitaine). In: *Proc. Colloque sur le Crétacé Inférieur, Lyon, Sept. 1963, Mém. Bur. Rech. géol. minières.* **34**, 669–694.

Setudehnia, A. (1978). The Mesozoic sequence in south-west Iran and adjacent areas. *J. Petrol. Geol.* **1** (1), 3–42.

Seyed-Enami, K. (1971a). The Jurassic Badamu Formation in the Kerman Region with remarks on the Jurassic stratigraphy of Iran. In: *Contributions to the Palaeontology and Stratigraphy of Iran, 1, Rep. geol. Surv. Iran.* **19**, 5–80.

Seyed-Enami, K. (1971b). A summary of the Triassic in Iran. In: *Contributions to the Palaeontology and Stratigraphy of Iran, 2, Rep. geol. Surv. Iran.* **20**, 41–53.

Seyed-Enami, K. (1975). Jurassic-Cretaceous boundary in Iran. *Bull. Am. Assoc. Petrol. Geol.* **59** (2), 231–238.

Seyfried, H. (1978). Der subbetische Jura von Murcia (Sudost-Spanien). *Geol. Jahrb.* **B29**, 3–201.

Sharief, F.A. (1983). Permian and Triassic geological history and tectonics of the Middle East. *J. Petrol. Geol.* **6** (1), 95–102.

Sheldon, R.P. (1964). Paleolatitudinal and paleogeographic distribution of phosphorite. *Prof. pap. U.S. geol. Surv.* **501-C**, 106–113.

Shepheard, W.W. and Hills, L.V. (1970). Depositional environments Bearpaw-Horseshoe Canyon (Upper Cretaceous) transition zone Drumheller 'Badlands', Alberta. *Bull. Can. Pet. Geol.* **18** (2), 166–215.

Sherwin, D.F. (1973). Scotian Shelf and Grand Banks. In: McCrossan, R.G. (ed.) *The future petroleum provinces of Canada – their geology and potential, Mem. Can. Soc. Petrol. Geol.* **1**, 519–559.

Sichler, B., Olivet, J-L., Auzende, J-M., Jonquet, H., Bonnin, J. and Bonifay, A. (1980). Mobility of Morocco. *Can. J. Earth Sci.* **17** (11), 1546–1558.

Simkjavicius, P.J., Grigjalis, A.A., Rotkyte, C.M., Monkevic, K.N., Moiseeva, T.I., Visnjakov, I.B., Glusko, U.V., Dulub, V.G., Pomjanovskaja, G.M., Sandler, J.M. and Romanov, L.F. (1983). Stratigraphie und Fazies des Bajoc, Bathan und Callov in den Sedimentationsbecken am W-Rand der Osteuropäischen Tafel. *Z. angew. Geol. DDR.* **29** (4), 157–166.

Simon, P. (1969). Paläogeographie der nordwestdeutschen Malm-Eisenerze. In: Bottke, H., Dengler, H., Finkenwirth, A. *et al. Sammelwerk Deutsche Eisenerzlagerstätten: II Eisenerze im Deckebirge (Postvaristikum) 1. Die Marin-sedimentaren Eisenerze des Jura in Nordwestdeutschland, Beih. geol. Jahrb.* **79**, 339–345.

Skarmeta, J. (1976). Evolución tectónica y paleogeográfica de los Andes Patagónicos de Aisén (Chile) durante el Neocomiano. In: *Proc. Primer Congr. geol. Chileno, Santiago, August 1976, 1, Sect. B Geotectonics.* **B-1-5.**

Skarpnes, O., Hamar, G.P., Jakobson, K.H. and Ormaasen, D.E. (1980). Regional Jurassic setting of the North Sea, north of the Central Highs. In: *The sedimentation of the North Sea Reservoir rocks, Proc. Symp. Geilo, May 1980. Nor. Pet. Foren.* **III**, 1–8.

Skwarko, S.K., Nicholl, R.S. and Campbell, K.S.W. (1976). The Late Jurassic mollusca, conodonts and brachiopods of the Kuta Formation and its Triassic palaeogeography, Papua New Guinea. *B.M.R. J. Aust. Geol. Geophys.* **1** (3), 219–230.

Slaczka, A. and Thompson, S. III. (1981). A revision of the flux-oturbidite concept based on type examples in the Polish Carpathian Flysch. *Ann. Soc. Geol. Pol.* **51** (1–2), 3–44.

Slowanska, B. and Bartys-Pelc, M. (eds.) (1976). Geology of Poland, 1, Stratigraphy (2) Mesozoic. *Wydawn. Geol. Warsaw.* 859pp.

Smart, J., Grimes, K.G. Doutch, H.F. and Pinchin, J. (1980). The Carpentaria and Karumba basins, north Queensland. *Bull. Bur. Min. Resour.* **202**, 73pp.

Smart, J. and Senior, B.R. (1980). Jurassic-Cretaceous basins of northeastern Australia. In: Henderson, R.A. and Stephenson, P.J. (eds.) *The geology and geophysics of northeastern Australia, Proc. 3rd Aust. geol. Conv., Townsville, 1978, Geol. Soc. Aust. (Inc. Queensland Div.).* 315–328.

Smith, C.I. (1970). Lower Cretaceous sedimentation and tectonics of the Coahuila and West Texas platforms. In: Seewald, K.O. and Sundeen, D. (eds.) *The geologic framework of the Chihuahua tectonic belt, a symposium in honor of Professor Ronald K. De Ford, Midland, Texas, Nov. 1970, Publ. West Texas Geol. Soc.* **71–59**, 75–82.

Smith, C.I. (1970). Lower Cretaceous stratigraphy, northern Coahuila, Mexico. *Rep. Invest. Bur. econ. Geol. Univ. Texas.* **65**, 101pp.

Smith, C.I. (1981). Review of geologic setting, stratigraphy and facies distribution of the Lower Cretaceous in northern Mexico. In: Katz, S.B. and Smith, C.I. (eds.) *Lower Cretaceous stratigraphy and structure, Northern Mexico. Field trip guidebook. West Texas geol. Soc. Publ.* **81-74**, 1–17.

Smith, E.G. and Warrington, G. (1971). The age and relationship of the Triassic rocks assigned to the lower part of the Keuper in north Nottinghamshire, north-west Lincolnshire and south Yorkshire. *Proc. Yorkshire geol. Soc.* **38** (2, 10), 201–227.

Smith, J.H.S., and Wennekers, J.H.N. (1977). Geology and hydrocarbon discoveries of Canadian Arctic Islands. *Bull. Am. Assoc. Petrol. Geol.* **61** (1), 1–27.

Smith, L.S. (1976). Paleoenvironments of the Upper Entrada Sandstone and the Curtis Formation on the west flank of the San Rafael Swell, Emery County, Utah. *Brigham Young Univ. Geol. Stud.* **23** (1), 113–171.

Soares, P.C. and Landim, P.M.B. (1976). Comparison between the tectonic evolution of the intracratonic and marginal basins in south Brazil. In: Almeida, F.F.M. (ed.) *Simposio internacional sobre as margens continentais de tipo Atlantico, Proc. Symp. Sao Paulo, October 1975, An. Acad. bras. Cienc.* **48**, Suppl., 3315–324.

Sohl, N.F. (1966). Upper Cretaceous gastropod assemblages of the Western Interior of the United States. In: Kauffmann, E.G. and Kent, H.C. (eds.) *Paleoenvironments of the Cretaceous seaway in the Western Interior, Prepr. Symp. Colorado School of Mines, Golden.* 1–38.

Sohl, N.F., Smith, C.C. and Christopher, R.A. (1976). Middle Cretaceous rocks of the Atlantic Seaboard and eastern Gulf Coastal plain of North America. In: *Evénements de la partie moyenne du Crétacé: rapports sur la biostratigraphie des régions clés, Uppsala 1975 – Nice 1976, Ann. Mus. Hist. Nice.* **4**, XXIV 1–12.

Sohn, I.G. (1967). Geographical implications of nonmarine Lower Cretaceous *Cyprideinae* in Israel, and *metacytheropteron parnesi,* n.sp. (ostracoda, crustacea). *Israel J. earth Sci.* **16** (3), 120–131.

Soliman, S.M. and El-Badry, O. (1970). Nature of Cretaceous sedimentation in western desert, Egypt. *Bull. Am. Assoc. Petrol. Geol.* **54** (12), 2349–2370.

Soquet, P., Peybernes, B., Bilote, M. and Debroas, E-J. (1977). La Chaîne Alpine des Pyrénées. *Géol. Alpine.* **53**, 193–216.

Sorgenfrei, T. (1969). Geological perspectives in the North Sea Area. *Medd. Dansk. geol. Foren.* **19** (2), 160–196.

Soto, F. (1979). Facies y ambientes deposicionales Cretácicos, área centro – sur de la Cuenca Marañón. In: Valdivia Ampuero, H., Díaz Bernal, N. and Megard, F. (eds.) *Anales del IV congreso peruano de geología, Lima, Aug. l978. Bol. Soc. geol. Perú.* **60**, 233–250.

Speed, R.C. (1978). Paleogeographic and plate tectonic evolution of the Early Mesozoic marine province of the western Great Basin. In: Howell, D.G. and McDougall, K.A. (eds.) *Mesozoic paleogeography of the western United States, Proc. 2nd Symp. Pacific Coast Paleogeogr., April 1978, Pacific Sect. Soc. econ. Paleontol. Mineral.* 253–270.

Speetzen, E., El-Arnauti, A. and Kaever, M. (1974). Beitrag zur Stratigraphie und Paläogeographie der Kreide-Basisschichten am S.O.-Rand der Westfälischen Kreidemulde (NW-Deutschland). *Neues Jahrb. Geol. Paläontol. Abhandlungen.* **145** (2), 207–241.

Sprigg, R.C. (1967). A short geological history of Australia. *Aust. Petrol. Explor. Assoc. J.* **7** (2), 59–82.

Springer, G.D., MacDonald, W.D. and Crockford, M.B.B. (1964). Jurassic. In: McGrossan, R.G. and Glaister, R.P. (eds.) *Geological history of Western Canada, Publ. Alberta Soc. Petrol. Geol., Calgary.* 137–155.

Stabler, C.L. and Marquez, D.B. (1977). Initiation of Lower Cretaceous reefs in Sabinas basin, northwest Mexico. In: Bebout, D.G. and Loucks, R.C. (eds.) *Cretaceous carbonates of Texas and Mexico: applications to subsurface exploration, Rep. Invest. Bur. econ. Geol. Univ.*

Texas. **89**, 299–301.

Stackelburg, U. von, Exon, N.F., Rad, U. von, Quilty, P., Shafik, S., Beiersdorf, H., Seibertz, E. and Veevers, J.J. (1980). Geology of the Exmouth and Wallaby Plateaus off northwest Australia: sampling of seismic sequences. *B.M.R. J. Aust. Geol. Geophys.* **5** (2), 113–140.

Stamm, R. and Thein, J. (1982). Sedimentation in the Atlas Gulf III: Turonian carbonates. In: Rad, U. von, Hinz, K. Sarnthein, M. and Seibold, E. (eds.) *Geology of the Northwest African Continental Margin.* Springer-Verlag, 459–474.

Stanley, G.D. Jnr. (1980). Triassic carbonate buildups of western North America: comparisons with the Alpine Triassic of Europe. In: Gaetani, M. (ed.) *Contributions to the Triassic Stratigraphy. Proc. Riccardo Assereto and Giulio Pisa field Symp. on Triassic stratigraphy of the Southern Alps, Bergamo, June 1979. Riv. Ital. Paleontol. Stratig.* **85** (3–4), 877–894.

Stanley, K.O., Jordan, W.M. and Dott, R.H. (1971). New hypothesis of early Jurassic paleogeography and sediment dispersal for western United States. *Bull. Am. Assoc. Petrol. Geol.* **55** (1), 10–19.

Stanley, R.G. (1981). Middle Jurassic shoaling of the central High Atlas sea near Rich, Morocco. *J. sediment. Petrol.* **51** (3), 895–907.

Stehli, F.G., Creath, W.B., Upshaw, C.F. and Forgotson, J.M. (1972). Depositional history of Gulfian Cretaceous of East Texas embayment. *Bull. Am. Assoc. Petrol. Geol.* **56** (1), 38–67.

Steinhauser, N. and Charollais, J. (1971). Observations nouvelles et réflexions sur la stratigraphie du 'Valanginien' de la région neuchâteloise et ses rapports avec le Jura méridional. *Géobios.* **4** (1), 7–59.

Steinitz, G. (1970). Chert 'dike' structures in Senonian chert beds, southern Negev, Israel. *J. sediment. Petrol.* **40** (4), 1241–1254.

Stelck, C.R. (1975a). The Upper *Miliammina manitobensis* Zone in northeastern British Columbia. In: Caldwell, W.G.E. (ed.) *The Cretaceous System in the Western Interior of North America. Proc. Int. Symp. Univ. Saskatchewan, Saskatoon, May 1973, Spec. Pap. geol. Assoc. Can.* **13**, 253–276.

Stelck, C.R. (1975b). Basement control of Cretaceous sand sequences in western Canada. In: Caldwell, W.G.E. (ed.) *The Cretaceous System in the Western Interior of North America. Proc. Int. Symp. Univ. Saskatchewan, Saskatoon, May 1973, Spec. Pap geol. Assoc. Can.* **13**, 427–440.

Stelck, C.R. and Kramers, J.W. (1980). Freboldiceras from the Grand Rapids Formation of north-Central Alberta. *Bull. Can. Pet. Geol.* **28** (4), 509–521.

Sterlenko, Yu.A. and Mosyakin, Yu.A. (1975). Features of geologic structure and formation of Paleozoic and early Mesozoic deposits of central and eastern Ciscaucasia with respect to oil and gas. *Int. geol. Rev.* **17** (9), 1013–1020.

Stets, J. and Wurster, P. (1981). Zur Structurgeschichte des Hohen Atlas in Marokko. *Geol. Rdsch.* **70** (3), 801–841.

Stevaux, J. (1971). Les facies du Keuper en Aquitaine; paléogéographie et dépendances avec leur substratum. *Bull. Cent. Rech. Pau.* **5** (2), 357–36l.

Stevaux, J. and Winnock, E. (1974). Les bassins du Trias et du Lias inférieur d'Aquitaine et leurs épisodes évaporitiques. *Bull. Soc. géol. Fr. Sér. 7.* **16** (6), 679–695.

Stevens, G.R. (1963). Faunal realms in Jurassic and Cretaceous belemnites. *Geol. Mag.* **100** (6), 481–497.

Stevens, G.R. (1980). Southwest Pacific faunal palaeobiogeography in

Mesozoic and Cenozoic times: a review. In: Ballance, P.F. (ed.) *Proc. Symp. S.W. Pacific paleobiogeography, ANZAAS Congr., Univ. Auckland Jan. 1979. Paleogeogr. Palaeoclimatol. Palaeoecol.* **31** (2–4), 153–196.

Stevens, G.R. and Speden, I.G. (1978). New Zealand. In: Moullade, M. and Nairn, A.E.M. (eds.) *The Phanerozoic geology of the world, 2, The Mesozoic, A.* Elsevier, Amsterdam, 251–328.

Stewart, J.H. Poole, F.G. and Wilson, R.F. (1972a). Stratigraphy and origin of the Chinle Formation and related Upper Triassic strata in the Colorado Plateau region. *Prof. pap. U.S. geol. Surv.* **690**, 336pp.

Stewart, J.H., Poole, F.G. and Wilson, R.F. (1972b). Stratigraphy and origin of the Triassic Moenkopi Formation and related strata in the Colorado Plateau region. *Prof. pap. U.S. geol. Surv.* **691**, 195pp.

Stibane, F.R. (1967). Paläogeographie und Tektogenese der Kolumbianischen Anden. *Geol. Rdsch.* **56** (2), 629–642.

Stipanicic, P.N. (1957). El Sistema Triásico en la Argentina. In: *20th Int. geol. Congr., Ciudad de México, 1956, Sect. II, El Mesozoico del hemisferio occidental y sus correlaciones Mundiales.* 73–112.

Stipanicic, P.N. (1969). Las sucesiones Triásicas Argentinas. *Proc. 1st Int. Symp. Gondwana Strat. Buenos Aires (1967), 2, UNESCO.* 1121–1141.

Stocklin, J. (1968). Salt deposits of the Middle East. In: Mattox, R.B. (ed.) *Saline deposits. Proc. Symp. Int. Confr., Houston, Texas, 1962, Spec. Pap. geol. Soc. Am.* **88**, 157–182.

Stoemer, N. and Wienholz, E. (1965). Mikrobiostratigraphie an der Lias/Doger Grenze in Bohrungen nördlich der mitteldeutschen Hauptscholle. *Jahrb. geol.* **1**, 533–571 (Publ. 1967).

Stokes, W.L. (1961). Fluvial and aeolian sandstone bodies in Colorado Plateau. In: Peterson, J.A. and Osmond, J.C. (eds.) *Geometry of Sandstone bodies. Am. Assoc. Petrol. Geol.* 151–178.

Stoneley, R. (1974). Evolution of the continental margins bounding a former southern Tethys. In: Burk, C.A. and Drake, C.L. (eds.) *The geology of continental margins.* Springer-Verlag, New York, 889–906.

Stott, D.F. (1962). Upper Cretaceous marine rocks of Rocky Mountain Foothills. *J. Alberta Soc. Pet. Geol.* **10** (5), 228–240.

Stott, D.F. (1968). Lower Cretaceous Bullhead and Fort St. John Groups, between Smoky and Peace Rivers, Rocky Mountain Foothills, Alberta and British Columbia. *Bull. geol. Surv. Can.* **152**, 279pp.

Stott, D.F. (1973). Lower Cretaceous Bullhead Group Between Bullmoose Mountain and Tetsa River, Rocky Mountain Foothills, northeastern British Columbia. *Bull. geol. Surv. Can.* **219**, 228pp.

Suarez, M. and Pettigrew, T.H. (1976). An upper Mesozoic island-arc-back-arc system in the southern Andes and South Georgia. *Geol. Mag.* **113** (4), 305–400.

Sun Dang-li and Zhang Bing-gao. (1983). Aspects of the marine Cretaceous of China. *Cret. Res.* **4** (2), 145–158.

Sundberg, F.A. (1980). Late Cretaceous paleoecology of the Holz Shale, Orange County, California. *J. Paleontol.* **54** (4), 840–857.

Surlyk, F. (1978a). Mesozoic geology and palaeogeography of Hochstetter Forland, East Greenland. *Bull. geol. Soc. Dan.* **27** (1), 73–87.

Surlyk, F. (1978b). Submarine fan sedimentation along fault scarps on tilted blocks (Jurassic-Cretaceous boundary, East Greenland). *Bull. Gronlands. geol. Unders.* **128**, 108pp.

Surlyk, F. and Clemmensen, L.B. (1975). A Valanginian turbidite sequence and its palaeogeographical setting (Kuhn O, East

Greenland). *Medd. Dansk geol. Foren.* **24** (1), 61–73.

Surlyk, F. and Clemmensen, L.B. (1983). Rift propagation and eustasy as controlling factors during Jurassic inshore and shelf sedimentation in north East Greenland. In: Boersma, J.R. and Terwindt, J.H.J. (eds.) *Basin analysis and sedimentary facies: sedimentology at various scales. Sediment. Geol.* **34** (2–3), 119–143.

Surlyk, F., Clemmensen, L.B. and Larsen, H.C. (1981). Post-Paleozoic evolution of the East Greenland continental margin. In: Kerr, J.W. and Fergusson, A.J. (eds.) *Geology of the North Atlantic Borderlands. Proc. Symp. St. John's, Newfoundland, June 1979, Mem. Can. Soc. Petrol. Geol.* **7**, 611–645.

Svoboda, J. *et al.* (1966). Regional geology of Czechoslovakia, 1, The Bohemian Massif. *Geol. Surv. Czechoslovakia, Czech. Acad. Sci., Prague* 668pp.

Swartz, D.H. and Arden, D.D. (1960). Geologic history of Red Sea area. *Bull. Am. Assoc. Petrol. Geol.* **44** (10), 1621–1637.

Szigeti, G.J. and Fox, J.E. (1981). Unkpapa Sandstone (Jurassic), Black Hills, South Dakota: an eolian facies of the Morrison Formation. In: Ethridge, F.G. and Flores, R.M. (ed.) *Recent and ancient non-marine depositional environments: models for exploration. Spec. Publ. Soc. econ. Paleontol. Mineral.* **31**, 331–349.

Szyperko-Sliwczynska, A. (1979). Trias dolny w potnocno-wschodneij Polsce. *Pr. Inst. geol. Warszawa.* **91**, 77pp.

Talbot, M.R. (1974). Ironstones in the Upper Oxfordian of southern England. *Sedimentol.* **21** (3), 433–450.

Tanner, P.W.G. (1982). Geologic evolution of South Georgia. In: Craddock, C. (ed.) *Antarctic geoscience. Proc. Symp. Antarctic geology and geophysics, Madison, Wisconsin Aug. 1977, Int. Union geol. Sci. Ser. B.* **4**, Univ. Wisconsin Press, 167–176.

Tanner, W.F. (1955). Paleogeographic reconstructions from cross-bedding studies. *Bull. Am. Assoc. Petrol. Geol.* **39** (12), 2471–2483.

Tanner, W.F. (1965). Upper Jurassic paleogeography of the Four Corners region. *J. sediment. Petrol.* **35** (3), 564–574.

Tardy, M. (1970). Etude géologique de la partie sud du massif du Monte Baldo (Alpes méridionales, province de Vérone, Italie). *Bull. Soc. géol. Fr. Sér. 7.* **12** (1), 78–87.

Tardy, M. (1977). Essai sur la reconstitution de l'évolution paléogéographique et structurale de la partie septentrionale du Mexique au cours du Mésozoïque et du Cénozoïque. *Bull. Soc. géol. Fr. Sér. 7.* **19** (6), 1297–1308.

Taylor, D.J. (1971). Foraminifera and the Cretaceous and Tertiary depositional history in the Otway Basin in Victoria. In: Wopfner, H. and Douglas, J.G. (eds.) *The Otway Basin of southeastern Australia. Spec. Bull. geol. Surv. South Aust.* 217–233.

Tchoumatchenco, P. (1972). Thanatocoenoses and biotopes of Lower Jurassic brachiopods in central and western Bulgaria. *Palaeogeogr. Palaeoclimatol. Palaeoecol.* **12** (4), 227–242.

Tempier, C. (1966). Les facies du Jurassique terminal dans les chaînes subalpines méridionales au Sud et à l'Est de la Durance. Leur répartition géographique. *Bull. Soc. géol. Fr. Sér. 7.* **8** (3), 468–470.

Tennyson, M.E. and Cole, M.R. (1978). Tectonic significance of Upper Mesozoic Methow-Pasayten sequence, northeastern Cascade Range, Washington and British Columbia. In: Howell, D.G. and McDougall, K.A. (eds.) *Mesozoic paleogeography of the western United States, Proc. 2nd Symp. Pacific Coast Paleogeogr. April 1978, Pacific Sect. Soc. econ. Paleontol. Mineral.* 499–508.

Termier, G. and Termier, H. (1981). Spongiaires du Crétacé moyen.

Cret. Res. **2**, 427–433.

Thiede, J. (1979). History of the North Atlantic Ocean: evolution of an asymmetric zonal paleoenvironment in a latitudinal ocean basin. In: Talwani, M., Hay, W. and Ryan, W.B.F. (eds.) *Deep drilling results in the Atlantic Ocean: continental margins and paleoenvironment. Proc. Symp. Harriman, New York, March 1978, Am. geophys. Union. Maurice Ewing Ser.* **3**, 275–296.

Thiermann, A. (1969). Zur Stratigraphie und Fazies der marinen Unterkreide im Gebiet des Ochtruper Sattels am Nordrand der Rheinischen Masse. *Fortschr. Geol. Rheinl. Westfalen* **17**, 219–242.

Thierry, J. (1976). Paléobiogéographie de quelques *Stephanoceratceae* (Ammonitina) du Jurassique Moyen et Supérieur; une confrontation avec la théorie mobiliste. *Géobios.* **9** (3), 291–322.

Thierry, J. (1982). Téthys, Mésogée et Atlantique au Jurassique: quelques reflexions basées sur les faunes d'Ammonites. In: *Proc. Symp. Biogéographie et tectonique des plaques, Brest, Mar. 1982. Bull. Soc. géol. Fr. Sér. 7.* **24** (5–6), 1053–1067.

Thies, D. (1979). Selachierzahne aus der nordwestdeutschen Unterkreide. In: Wiedmann, J. (ed.) *Aspekte der Kreide Europas, Proc. 1st Symp. German Cretaceous, Münster, April 1978, Int. Union geol. Sci., Ser. A.* **6**, E. Schweizerbart'sche Verlagsbuchhandlung, Stuttgart, 211–222.

Thomas, W.A. and Mann, C.J. (1966). Late Jurassic depositional environments, Louisiana and Arkansas. *Bull. Am. Assoc. Petrol. Geol.* **50** (1), 178–182.

Thompson, J.E. (1967). A geology history of eastern New Guinea. *Aust. Petrol. Explor. Assoc. J.* **7** (2), 83–93.

Thomson, M.R.A. (1982). Mesozoic paleogeography of West Antarctica. In: Craddock, C. (ed.) *Antarctic Geoscience, Proc. 3rd Symp. Antarctic geology and geophysics, Madison, Wisconsin Aug. 1977, Int. Union geol. Sci. Publ. Ser. B.* **4**, Univ. Wisconsin Press, 331–337.

Thurrow, J., Kuhnt, W. and Wiedmann, J. (1982). Zeitlicher und paläogeographischer Rahmen der Phthanit-und Black Shale – Sedimentation in Marokko. In: Wiedmann, J., Kullmann, J. and Schonenberg, R. (eds.) *Subsidenz-Entwicklung im Kantabrischen Variszkum und an passiven Kontinentalrändern der Kreide, 2, Kreide. Neues Jahrb. Geol. Paläontol. Abhandlungen.* **165** (1), 147–176.

Tintori, A. (1977). Toarcian fishes from the Lombardian basin. *Boll. Soc. Paleontol. Ital.* **16** (2), 143-152.

Tipper, H.W. and Richards, T.A. (1976). Jurassic stratigraphy and history of north-central British Columbia. *Bull. geol. Surv. Can.* **270**, 73pp.

Tischer, G. (1965). Über die Wealden-Ablagerung und die Tektonik der östlichen Sierra de los Cameros in den nordwestlichen Iberischen Ketten (Spanien). In: Beuther, A., Dahm, H., Kneuper-Haack, F. *et al.* (1966) *Der Jura und Wealden in Nordost-Spanien. Beih. geol. Jahrb.* **44**, 123–164.

Tissot, B., Deroo, G. and Herbin, J.P. (1979). Organic matter in Cretaceous sediments of the North Atlantic: contribution to sedimentology and paleogeography.In: Talwani, M., Hay, W. and Ryan, W.B.F. (eds.) *Deep drilling results in the Atlantic Ocean: continental margins and paleoenvironment. Proc. Symp. Harriman, New York, March 1978, Am. Geophys. Union, Maurice Ewing Ser.,* **3**, 362–374.

Tomás, M.J.A. (1971). Consideraciones generales sobre la sedimentación y paleogeografía del Albense de la Cuenca Cantábrica. *Estud. geol. Inst. Invest. geol. Lucas Mallada.* **27** (4), 325–334.

Tourtelot, H.A. (1962). Preliminary investigation of the geologic setting and chemical composition of the Pierre Shale, Great Plains Region. *Prof. pap. U.S. geol. Surv.* **390**, 74pp.

Tourtelot, H.A. and Cobban, W.A. (1968). Stratigraphic significance and petrology of phosphate nodules at base of Niobrara Formation, east flank of Black Hills, South Dakota. *Prof. pap. U.S. geol. Surv.* **594-L**, 22pp.

Tourtelot, H.A. and Rye, R.O. (1969). Distribution of oxygen and carbon isotopes in fossils of late Cretaceous age, Western Interior region of North America. *Bull. geol. Soc. Am.* **80** (10), 1903–1922.

Toutin, N. and Lefranc, J-P. (1970). Le Turonien du Nord du Tademait (Sahara central, Algérie). Etude stratigraphique et sédimentologique. *Bull. Soc. géol. Fr. Sér. 7.* **12** (3), 435–446.

Tozer, E.T. (1960). Summary account of Mesozoic and Tertiary stratigraphy, Canadian Arctic Archipelago. In: Raasch, G.O. (ed.) *Geology of the Arctic,* **1**, *Proc. 1st Int. Symp. Arctic geology, Alberta, 1960,* Univ. Toronto Press. 381–402.

Tozer, E.T. (1982). Marine Triassic faunas of North America: their significance for assessing plate and terrain movements. In: *Proc. Symp. Triassic global geology, Wurzburg, Feb. (1982), Geol. Rdsch.* **71** (3), 1077–1104.

Trammer, J. (1980). The isochronous synsedimentary movements at the Anisian/Ladinian boundary in the Muschelkalk Basin and the Alps. In: Gaetani, M. (ed.) *Contributions to the Triassic stratigraphy, Proc. Riccardo Assereto and Giulio Pisa field Symp. on Triassic stratigraphy of the Southern Alps, Bergamo, June 1979. Riv. Ital. Paleontol. Stratigr.* **85** (3–4), 931–936.

Trevisan, L. (1963). La paléogéographie du Trias de l'Apennin septentrional et central, et ses rapports avec la tectogénèse. In: Durand-Delga, M. (ed) *Livre à la Mémoire du Professeur Paul Fallot: L'évolution paléogéographique et structurale des domaines méditerranéens et Alpins d'Europe. Mém. Soc. géol. Fr. Hors Sér. 1,* **2**, 217–226.

Troger, K-A. (1968). Zur Bedeutung ökologischer Faktoren für die Leitfossilien. *Geologie. DDR.* **17** (1), 68–75.

Troger, K-A. (1971). Gesetzmässigkeiten der Beziehung von Bio – und Lithogenese im litoralen und neritischen Bereich von Karbonatsedimenten (Bereich Oberkreide). *Ber. Deutsch. Ges. geol. Wiss., Ser. A, Geol. Paläontol.* **16** (3–5), 447–464.

Troger, K-A. (1975). Fazies – Biofazies am Beispiel der Sächsischen und subherzynen Kreide. *Z. geol. Wiss. DDR.* **3** (10), 1265–1277.

Troger, K-A. (1978). Probleme der Paläontologie, Biostratigraphie und Paläobiogeographie Oberkretazischer Faunen (Cenoman-Turon)Westeuropas und der Russischen Tafel. *Z. geol. Wiss. DDR.* **6** (5), 557–570.

Troger, K-A. (1981a). Zu einigen Problemen der Alb-Cenoman-Transgression in Europa. *Z. angew. Geol. DDR.* **27** (4), 160-163.

Troger, K-A. (1981b). German Democratic Republic. In: Reyment, R.A. and Bengtson, P. (eds.) *Aspects of Mid-Cretaceous Regional Geology.* Academic Press, London, 1–28.

Troger, K.-A. and Kurze, M. (1980). Zur paläogeographischen Entwicklung des Mesozoikums im Südteil des Subjerzynen Beckens. *Z. geol. Wiss. DDR.* **8** (10), 1247–1265.

Trumpy, R. (1960). Paleotectonic evolution of the central and western Alps. *Bull. geol. Soc. Am.* **71** (6), 843–908.

Trumpy, R. (1971). Sur le Jurassique de la zone helvétique en Suisse. In: Vegh-Neubrandt, E. (ed.) *Colloque du Jurassique Méditerranéen, Budapest, Sept. 1969. Ann. Inst. geol. publici Hung.* **54** (2), 369–382.

Tucker, D.R. (1968). Lower Cretaceous geology, northwestern

Karnes County, Texas. *Bull. Am. Assoc. Petrol. Geol.* **52** (5), 820–851.

Turnsek, D., Buser, S. and Ogorelec, B. (1981). An Upper Jurassic reef complex from Slovenia, Yugoslavia. In: Toomey, D.F. (ed.) *European fossil reef models, Spec. Publ. Soc. econ. Paleontol. Mineral.* **30**, 361–369.

Tyler, N. and Ethridge, F.G. (1983). Depositional setting of the Salt Wash Member of the Morrison Formation, southwest Colorado. *J. Sediment. Petrol.* **53** (1), 67–82.

Urlichs, M. (1966). Zur Fossilführung und Genese des Feuerlettens, der Rat-Lias-Grenzschichten und des Unteren Lias bei Nürnberg. *Erlanger geol. Abhandlungen.* **64**, 42pp.

Urlichs, M. (1977). The Lower Jurassic in southwestern Germany. *Beitr. Naturkd. Ser. B.* **24**, 30pp.

Valecka, J. (1979). Paleogeografie a litofacialni vyvoj severozapdni casti ceski kridove panve. *Sb. geol. Ved. Praha.* **33** (1), 47–81.

Vallier, T.L., Underwood, M.B., Jones, D.L. and Gardner, J.V. (1980). Petrography and geologic significance of Upper Jurassic rocks dredged near Pribilof Island, southern Bering Sea continental shelf. *Bull. Am. Assoc. Petrol. Geol.* **64** (6), 945–950.

van Andel, Tj.H. (1958). Origin and classification of Cretaceous Paleocene and Eocene sandstones of western Venezuela. *Bull. Am. Assoc. Petrol. Geol.* **42** (4), 734–763.

Vavilov, N.N., Zinchenko, V.N., Kirina, T.L., Polubotko, I.V. and Repin, Y.S. (1980). Position of the Triassic-Jurassic boundary in the region west of Verkhoyansk. *Dokl. Acad. sci. U.S.S.R.* **251**, 54–56.

Veevers, J.J. (1969). Palaeogeography of the Timor Sea region. *Palaeogeogr. Palaeoclimatol. Palaeoecol.* **6** (2), 125–140.

Veevers, J.J. and Cotterill, D. (1978). Western margin of Australia: evolution of a rifted arch system. *Bull. geol. Soc. Am.* **89** (3), 337–355.

Veevers, J.J., Jones, J.G. and Powell, C.Mc.A. (1982). Tectonic framework of Australia's sedimentary basins. *Aust. Petrol. Explor. Assoc. J.* **22** (1), 283–300.

Veevers, J.J. and Rundle, A.S. (1979). Channel Country fluvial sands and associated facies of central-eastern Australia: modern analogues of Mesozoic desert sands of South America. *Palaeogeogr. Palaeoclimatol. Palaeoecol.* **26** (1), 1–16.

Veizer, J. (1970). Zonal arrangement of the Triassic rocks of the western Carpathians: a contribution to the dolomite problem. *J. sediment. Petrol.* **40** (4), 1287–1301.

Verma, K.K. (1968). Bagh Beds – their fauna and affinities with South Indian Cretaceous Formations. In: *Cretaceous-Tertiary Formations of South India, Proc. Seminar Bangalore, June 1966, Mem. geol. Soc. India.* **2**, 239–247.

Viel, G. (1979a). Litostratigrafia Ladinica: una revisione. Ricostruzione paleogeografica e paleostrutturale dell'area Dolomitico-Cadorina (Alpi Meridionali). 1. *Riv. Ital. paleontol. stratigr.* **85** (1), 85–125.

Viel, G. (1979b). Litostratigrafia Ladinica: una revisione. Ricostruzione paleografica e paleostrutturale dell'area Dolomitico-Cadorina (Alpi Meridionali). 2. *Riv. Ital. Paleontol. stratigr.* **85** (2), 297–352.

Villain, J-M. (1977). Le Maastrichtien dans sa région type (Limbourg, Pays-Bas). Etude stratigraphique et micropaléontologique. *Palaeontographica, Abt. A.* **157** (1–3), 1–87.

Viniegra, O.F. (1971). Age and evolution of salt basins of southeastern Mexico. *Bull. Am. Assoc. Petrol. Geol.* **55** (3), 478–494.

Viniegra, O.F. (1981). Great carbonate bank of Yucatan, southern Mexico. *J. Petrol. Geol.* **3** (3), 247–278.

Vinogradov, C. and Dragastan, O. (1975). Microfacial study of the Upper Jurassic and Lower Cretaceous deposits from the central part of the Moesic platform (Romania). *Rev. Roumaine. Geol. Geophys. Geogr.* **19**, 105–117.

Vinson, G.L. (1962). Upper Cretaceous and Tertiary stratigraphy of Guatemala. *Bull. Am. Assoc. Petrol. Geol.* **46** (4), 425–456.

Viohl, G. (1969). Die Keuper-Lias-Grenze in Sudfranken. *Erlanger geol. Abhandlungen.* **76**, 81pp.

Virgili, C. (1962). Le Trias du Nord-Est de l'Espagne. In: Durand-Delga, M. (ed.) *Livre à la Mémoire du Professeur Paul Fallot: L'évolution paléogéographique et structurale des domaines méditerranéens et Alpins d'Europe. Mém. Soc. géol. Fr., Hors Sér. 1.* **1**, 301–312.

Virgili, C., Suarez-Vega, L.C. and Rincon, R. (1971). Le Mésozoïque des Asturies (Nord de l'Espagne). In: *Proc. Symp. Histoire structural du Golfe de Gascogne, Rueil-Malmaison, Déc. 1970, Publ. Inst. Fr. Pét., Collection Colloq. et Séminaires.* **22** (2), Editions Technip, Paris, V-4, 1–20.

Visscher, H. and Zwan, C.J. van der (1981). Palynology of the circum-Mediterranean Triassic: phytogeographical and palaeoclimatological implications. *Geol. Rdsch.* **70** (2), 625–634.

Voigt, E. (1965). Zur Temperaturkurve der Oberen Kreide in Europa. In: *Proc. Symp. 'Palaoklima', March 1964, Köln, Geol. Rdsch.* **54** (1), 270–317.

Volkheimer, W. (1969–70). Neuere Ergebnisse der Anden-Stratigraphie von Süd-Mendoza (Argentinien) und benachbarter Gebiete und Bemerkungen zur Klimageschichte des südlichen Andenraums. *Geol. Rdsch.* **59** (3), 1088–1124.

Waldeck, H. (1969). Gefügeuntersuchungen in Wealdenkonglomeraten des Osterwaldes und Nesselberges südlich Hannover. *Geol. Jahrb.* **87**, 229–276.

Wall, J.H. (1975). Diatoms and radiolarians from the Cretaceous System of Alberta – a preliminary report. In: Caldwell, W.G.E. (ed.) *The Cretaceous System in the Western Interior of North America. Proc. Int. Symp. Univ. Saskatchewan, Saskatoon, May 1973, Spec. Pap. geol. Assoc. Can.* **13**, 391–410.

Wall, J.H. and Singh, C. (1975). A late Cretaceous microfossil assemblage from the Buffalo Head Hills, north-central Alberta. *Can. J. Earth Sci.* **12** (7), 1157–1174.

Wall, J.H., Sweet, A.R. and Hills, L.V. (1971). Paleoecology of the Bearpaw and contiguous Upper Cretaceous Formations in the C.P.O.G. Strathmore Well, southern Alberta. *Bull. Can. Pet. Geol.* **19** (3), 691–702.

Walley, C.D. (1983). The palaeoecology of the Callovian and Oxfordian strata of Majdal Shams (Syria) and its implications for Levantine palaeogeography and tectonics. *Palaeogeogr. Palaeoclimatol. Palaeoecol.* **42** (3–4), 323–340.

Walper, J.L., Henk, F.H., Loudon, E.J. and Raschilla, S.N. (1979). Sedimentation on a trailing plate margin: the northern Gulf of Mexico. *Trans. Gulf Coast Assoc. geol. Soc.* **29**, 188–201.

Walther, H.W. (1982). Zur Bildung von Erz-und Minerallagerstätten in der Trias von Mitteleuropa. In: *Proc. Symp. Triassic global geology, Würzburg, Feb. 1982, Geol. Rdsch.* **71** (3), 835–855.

Ward, C.R. (1980). Notes on the Bulgo Sandstone and the Bald Hill Claystone. In: Herbert, C. and Helby, R. (eds.) *A guide to the Sydney Basin, Bull. geol. Surv. New South Wales.* **26**, 178–187.

Bibliography

Wardlaw, N.C. and Nicholls, G.D. (1972). Cretaceous evaporites of Brazil and West Africa and their bearing on the theory of continent separation. In: *Proc. 24th Int. geol. Congr., Ottawa 19, Sect. 6, Stratigraphy and sedimentology.* 43–55.

Ware, M. and Windle, T.M.F. (1981). Micropalaeontological evidence for land near Cirencester, England, in Forest Marble (Bathonian) times: a preliminary account. *Geol. Mag.* 118 (4), 415–420.

Warner, D.L. (1964). Mancos - Meseverde (Upper Cretaceous) intertonguing relations southeast Piceance basin, Colorado. *Bull. Am. Assoc. Petrol. Geol.* 48 (7), 1091–1107.

Warrington, G. (1974). Les évaporites du Trias britannique. *Bull. Soc. géol. Fr. Sér. 7.* 16 (6), 708–723.

Warris, B.J. (1973). Plate tectonics and the evolution of the Timor Sea, northwest Australia. *Aust. Petrol. Explor. Assoc. J.* 13 (1), 13–18.

Waters, J.A., McFarland, P.W. and Lea, J.W. (1955). Geologic framework of Gulf Coastal Plain of Texas. *Bull. Am. Assoc. Petrol. Geol.* 39 (9), 1821–1850.

Weddle, T.K. and Hubert, J.F. (1983). Petrology of Upper Triassic sandstones of the Newark Supergroup in the northern Newark, Pomperaug, Hartford and Deerfield basins. *Northeastern Geol.* 5 (1), 8–22.

Weichman, B.E., Glaze, R.E., Keller, E.R. *et al.* (1965). Geologic history of Powder River basin. *Bull. Am. Assoc. Petrol. Geol.* 49 (11), 1893–1907.

Weimer, R.J. (1960). Upper Cretaceous stratigraphy, Rocky Mountain area. *Bull. Am. Assoc. Petrol. Geol.* 44 (1), 1–20.

Weimer, R.J. (1961). Spatial dimensions of Upper Cretaceous sandstones, Rocky Mountain area. In: Peterson, J.A. and Osmond, J.C. (eds.) *Geometry of Sandstone bodies, Am. Assoc. Petrol. Geol.* 82–97.

Weimer, R.J. (1966). Time-stratigraphic analysis and petroleum accumulations, Patrick Draw field, Sweetwater County, Wyoming. *Bull. Am. Assoc. Petrol. Geol.* 50 (10), 2150–2175.

Weimer, R.J. (1983). Relation of unconformities, tectonics and sea level changes, Cretaceous of the Denver Basin and adjacent areas. In: Reynolds, M.W. and Dolly, E.D. (eds.) *Mesozoic Paleogeography of the West-Central United States, Proc. 2nd Rocky Mt. Paleogeogr. Symp., Denver, Colorado. Rocky Mt. Sect., Soc. econ. Paleontol. Mineral.* 359–376.

Weimer, R.J. and Land, C.B. (1975). Maestrichtian deltaic and interdeltaic sedimentation in the Rocky Mountain region of the United States. In: Caldwell, W.G.E. (ed.) *The Cretaceous System in the Western Interior of North America. Proc. Int. Symp. Univ. Saskatchewan, Saskatoon, May 1973, Spec. Pap. geol. Assoc. Can.* 13, 633–666.

Weippert, D. von (1968). Über kretazische Sedimente im Nördlichen Hindukush-Vorland (Nord-Afghanistan). *Z. Deutsch. geol. Ges.* 117 (1965), (2–3), 829–854.

Weiss, M.P. (1969). Oncolites, paleoecology, and laramide tectonics, central Utah. *Bull. Am. Assoc. Petrol. Geol.* 53 (5), 1105–1120.

Welzol, E. (1968). Formation und Fazies des fränkischen Domeriums. *Erlanger geol. Abhandlungen.* 69, 86pp.

Wendt, J. (1968–1969). Die stratigraphisch-paläogeographische Entwicklung des Jura in Westsizilien. *Geol. Rdsch.* 58 (3), 735–755.

West, I.M. (1975). Evaporites and associated sediments of the basal Purbeck Formation (Upper Jurassic) of Dorset. *Proc. Geol. Assoc. Lond.* 86 (2), 205–225.

Westermann, G. (1958). Ammoniten-fauna und Stratigraphie des Bathonien N.W. Deutschlands. *Beih. geol. Jahrb.* 32, 103pp.

Weyl, R. (1965). Die Paläogeographische Entwicklung des Mittelamerikanischen-Westindischen Raumes. *Geol. Rdsch.* 54 (2), 1213–1240.

Weyl, R. (1980). Geology of Central America. 2nd Edn. *Beitrage zur regionalen geologie der Erde,* 15, Gebruder Borntraeger, Berlin, 371pp.

Whiteman, A. (1982). *Nigeria: its petroleum geology, resources and potential,* 1, *stratigraphy.* Graham and Trotman, London, 166pp.

Wiedmann, J. (1962). Contribution à la paléogéographie du Crétacé vascogotique et celtibérique septentrional (Espagne). In: Durand-Delga, M. (ed.) *Livre à la Mémoire du Professeur Paul Fallot: L'évolution paléogéographique et structurale des domaines méditerranéens et Alpins d'Europe. Mém. Soc. géol. Fr. Hors Sér.* 1, 1, 351–366.

Wiedmann, J. (1979). Itinéraire géologique à travers le Crétacé Moyen des chaînes vascogotiques et celtibériques (Espagne du nord). *Proc. Mid Cretaceous Events. Iberian Field Conf. Madrid, 1977, Guide II, Cuad. Geol. Iberica.* 5, 127–214.

Wiedmann, J. (1980). Paläogeographie und Stratigraphie im Grenzbereich Jura/Kreide Südamerikas. *Münster Forsch. Geol. Paläontol.* 51, 27–61.

Wiedmann, J. (1982). Grundzüge der kretazischen Subsidenz- Entwicklung im Südatlantik, in Marokko, Nordspanien und im Helvetikum. In: Wiedmann, J., Kullman, J. and Schonenberg, R. (eds.) *Subsidenz-Entwicklung im Kantabrischen Variszikum und an passiven Kontinentalrändern der Kreide, 2, Kreide. Neues Jahrb. Geol. Paläont. Abhandlungen.* 165 (1), 5–31.

Wiedmann, J., Bergner, H.D. and Labude, C. (1982). Becken-Evolution und Jungkimmerische Bewegungen an der Jura/Kreide-Grenze Marokkos. In: Wiedmann, J., Kullman, J. and Schonenberg, R. (eds.) *Subsidenz-Entwicklung im Kantabrischen Variszikum und an passiven Kontinentalrändern der Kreide, 2, Kreide, Neues Jahrb. Geol. Paläont. Abhandlungen.* 165 (1), 46–59.

Wiedmann, J., Butt, A. and Einsele, G. (1978). Vergleich von marokkanischen Kreide-Küstenaufschlüssen und Tiefseebohrungen (DSDP): Stratigraphie, Paläoenvironment und Subsidenz an einem passiven Kontinentalrand. *Geol. Rdsch.* 67 (2), 454–508.

Wienholz, R. (19XX) Geologischer Bau des Untergrundes im nordostdeutschen Flachland. *Jahrb. Geol.* 1 (1), 1–87.

Wildi, W. (1979). Evolution de la plate-forme carbonatée de type austro-alpin de la Dorsale calcaire (Rif interne, Maroc septentrional) au Mésozoïque. *Bull. Soc. géol. Fr. Sér. 7.* 21 (1), 49–56.

Wildi, W. (1981). Le Ferrysch: cône de sédimentation détritique en eau profonde à la bordure nord-ouest de l'Afrique au Jurassique Moyen à Supérieur (Rif externe, Maroc). *Eclog. geol. Helv.* 74 (2), 481–527.

Wildi, W. (1983). La chaîne tello-rifaine (Algérie, Maroc, Tunisie): structure, stratigraphie et évolution du Trias au Miocène. *Rev. Géol. phys. Géogr. dyn.* 24 (3), 201–297.

Wilkinson, A.F. (1982). Exploration for phosphate in Ecuador. *Trans. Instn. Ming. Metall., B. Appl. Earth Sci.* B91 (3), 130–145.

Will, H-J. (19XX). Untersuchungen zur Stratigraphie und Genese des Oberkeupers in Nordwestdeutschland. *Beih. geol. Jahrb.* 54, 240pp.

Williams, G.D. (1966). The Great Artesian Basin – origin and history. *Aust. Petrol. Explor. Assoc. J., Ann. Conf. Issue.* 85–92.

Williams, G.D. and Burk, C.F.Jnr. (1964). Upper Cretaceous. In: McGrossan, R.G. and Glaister, R.P. (eds.) *Geological history of Western Canada, Publ. Alberta Soc. Petrol. Geol., Calgary.* 169–189.

Williams, G.D. and Stelck, C.R. (1975). Speculations on the Cretaceous palaeogeography of North America. In: Caldwell, W.G.E. (ed.) *The Cretaceous System in the Western Interior of North America. Proc. Inst. Symp. Univ. Saskatchewan, Saskatoon, May 1973, Spec. Pap. geol. Assoc. Can.* **13**, 1–20.

Wills, L.J. (1976). The Trias of Worcestershire and Warwickshire. *Rep. Inst. geol. Sci. Lond.* **76** (2), 209pp.

Wilson, H.H. (1969). Late Cretaceous eugeosynclinal sedimentation, gravity tectonics, and ophiolite emplacement in Oman Mountains, southeast Arabia. *Bull. Am. Assoc. Petrol. Geol.* **53** (3), 626–671.

Wilson, H.H. (1974). Cretaceous sedimentation and orogeny in nuclear Central America. *Bull. Am. Assoc. Petrol. Geol.* **58** (7), 1348–1396.

Wilson, J.A. (1970). Vertebrate biostratigraphy of Trans Pecos Texas. In: Seewald, K.O. and Sundeen, D. (eds.) *The geologic framework of the Chihuahua tectonic belt, a symposium in honor of Professor Ronald K. De Ford, Midland, Texas, Nov. 1970, Publ. West Texas Geol. Soc.* **71–59**, 159–166.

Wilson, J.J. (1963). Cretaceous stratigraphy of central Andes of Peru. *Bull. Am. Assoc. Petrol. Geol.* **47** (1), 1–34.

Wilson, J.L. (1975). *Carbonate facies in geologic history.* Springer-Verlag, New York, 471pp.

Wilson, J.L. (1981). Lower Cretaceous stratigraphy in the Monterrey-Saltillo area. In: Katz, S.B. and Smith, C.I. (eds.) *Lower Cretaceous stratigraphy and structure, northern Mexico. Field trip guidebook. West Texas geol. Soc. Publ.* **81–74**, 78–84.

Wilson, L.M. (1981). Circum-North Atlantic tectono-stratigraphic reconstruction. In: Kerr, J.W. and Fergusson, A. (eds.) *Geology of the North Atlantic borderlands, Proc. Symp. St. John's Newfoundland, June 1979, Mem. Can. Soc. Petrol. Geol.* **7**, 167–184.

Wilson, M.D. (1970). Upper Cretaceous-Paleocene synorogenic conglomerates of southwestern Montana. *Bull. Am. Assoc. Petrol. Geol.* **54** (10), 1843–1867.

Wilson, R.C.L. (1968). Upper Oxfordian palaeogeography of southern England. *Palaeogeogr. Palaeoclimatol. Palaeoecol.* **4** (1), 5-28.

Wilson, R.C.L. (1975). Atlantic opening and Mesozoic margin basins of Iberia. *Earth planet. Sci. Lett.* **25** (1), 33–43.

Wilson, R.C.L. (1979). A reconnaissance study of Upper Jurassic sediments of the Lusitanian Basin. *Cienc. Terra (Fac. Cienc. Tec., Univ. Nova, Lisboa).* **5**, 53–84.

Wind, F.H. (1979). Maestrichtian-Campanian nanofloral provinces of the southern Atlantic and Indian Oceans. In: Talwani, M., Hay, W. and Ryan, W.B.F. (eds.) *Deep drilling results in the Atlantic Ocean: continental margins and paleoenvironment. Proc. Symp. Harriman, New York, 1978. Am. Geophys. Union, Maurice Ewing Ser.* **3**, 123–137.

Winn, R.D. (1978). Upper Mesozoic flysch of Tierra del Fuego and South Georgia Island: A sedimentologic approach to lithosphere plate restoration. *Bull. geol. Soc. Am.* **89** (4), 533–547.

Winnock, E. (1971). Geologie succincte du bassin d'Aquitaine (Contribution à l'histoire du Golfe de Gascogne). In: *Proc. Symp. Histoire structural du Golfe de Gascogne, Rueil-Malmasion, Déc. 1970, Publ. Inst. Fr. Pét., Collection Colloq. et Séminaires.* **22** (l), Editions Technip, Paris, IV-l, 1–30.

Winnock, E. (1973). Exposé succinct de l'évolution paléogéologique de l'Aquitaine. *Bull. Soc. géol. Fr. Sér. 7.* **15** (1), 5–12.

Winterer, E.L. and Bosellini, A. (1981). Subsidence and sedimentation on Jurassic passive continental margin, southern Alps, Italy. *Bull.*

Am. Assoc. Petrol. Geol. **65** (3), 394–421.

Wirth, W. (1964). Über zwei Unterkreiderelikte im nördlichen Sauerland. In: Die Kreide Westfalens – ein Symposium, *Fortschr. Geol. Reinl. Westfalen.* **7**, 403–420.

Witze, B.J., Ludvrgsen, G.A., Poppe, J.R. and Ravn, R.L. (1983). Cretaceous paleogeography along the eastern margin of the Western Interior Seaway, Iowa, Southern Minnesota, and eastern Nebraska and South Dakota. In: Reynolds, M.W. and Dolly, E.D. (eds.) *Mesozoic paleogeography of the West-Central United States. Proc. 2nd Rocky Mt. Paleogeogr. Symp., Denver, Colorado. Rocky Mt. Sect. Soc. econ. Paleontol. Mineral.* 225–252.

Wolburg, J. (1961). Sedimentations – Zyklen und Stratigraphie des Buntsandsteins in N.W. Deutschland. *Geotekton. Forsch.* **14** (1), 7–74.

Wolfart, R. (1967). Geologie von Syrien und dem Libanon. *Beiträge zur regionalen Geologie der Erde.* **6**, Gebrüder Borntraeger, Berlin, 326pp.

Wolfart, R. (1968). Tektonik und paläogeographische Entwicklung des mobilen Schelfes im Bereich von Syrien und dem Libanon. *Z. Deutsch. geol. Ges.* **117** (1965), (2–3), 544–589.

Wolfart, R. and Wittekindt, H. (1980). Geologie von Afghanistan. *Beiträge zur regionalen Geologie der Erde.* **14**, Gebrüder Borntraeger, Berlin, 500pp.

Wopfner, H., Kenley, P.R. and Thornton, R.C.N. (1971). Hydrocarbon occurrences and potential of the Otway Basin. In: Wopfner, H. and Douglas, J.G. (eds.) *The Otway Basin of southeastern Australia. Spec. Bull. geol. Surv. South Aust. Victoria.* 385–435.

Wormbs, J. (1976). Zur Stratigraphie und Biofazies des Doggers im Nordteil der DDR auf der Grundlage der Makrofauna. *Jahrb. Geol.* **5/6** (1969/1970), 329–365.

Worssam, B.C. (1978). The stratigraphy of the Weald Clay. *Rep. Inst. geol. Sci. Lond.* **78** (11), 23pp.

Wright, J.K. (1977). The Cornbrash Formation (Callovian) in North Yorkshire and Cleveland. *Proc. Yorkshire geol. Soc.* **41** (3, 27), 325–346.

Wright, R.P. (1973). Marine Jurassic of Wyoming and South Dakota: Its paleoenvironments and paleobiogeography. *Pap. Paleontol., Mus. Paleontol. Univ. Michigan.* **2**, 49pp.

Wulf, G.R. (1962). Lower Cretaceous Albian rocks in northern Great Plains. *Bull. Am. Assoc. Petrol. Geol.* **46** (8), 1371–1415.

Wurster, P. (1964). Geologie des Schilfsandsteins. *Mitt. geol. Staatsinst. Hamburg.* **33**, 140pp.

Wurster, P. (1968). Paläogeographie der deutschen Trias und die paläogeographische Orientierung der Lettenkohle in Südwestdeutschland. *Eclog. geol. Helv.* **61** (1), 157–166.

Wurster, P. and Stets, J. (1982). Sedimentation in the Atlas Gulf II: Mid Cretaceous events. In: Rad, U. von, Hinz, K., Sarnthein, M. and Seibold, E. (eds.) *Geology of the Northwest African Continental Margin.* Springer-Verlag, 439–458.

Yegorova, I.N., Popov, N.N. and Yakobson, G.P. (1975). Paleohydrogeologic conditions of oil and gas occurrences in Jurassic deposits of Ciscaucasia. *Int. geol. Rev.* **17** (8), 874–880.

Yorath, C.J. and Cook, D.G. (1981). Cretaceous and Tertiary stratigraphy and paleogeography, Northern Interior Plains, district of Mackenzie. *Mem. geol. Surv. Can.* **398**, 76pp.

Young, F.G. (1975). Upper Cretaceous stratigraphy, Yukon coastal plain and northwestern Mackenzie delta. *Bull. geol. Surv. Can.* **249**, 83pp.

Young, F.G., Myhr, D.W. and Yorath, C.J. (1976). Geology of the Beaufort-Mackenzie Basin. *Pap. geol. Surv. Can.* **76–11**, 63pp.

Young, G.A., Bellizzia, A., Renz, H.H., Johnson, F.W., Robie, R.H. and Mas Vall, J. (1956). Geologia de las cuencas sedimentarias de Venezuela y de sus campos petroliferos. *Publ. esp. Bol. Geol. Caracas.* **2**, 140pp.

Young, K. (1977). Middle Cretaceous rocks of Mexico and Texas. In: Bebout, D.G. and Loucks, R.G. (eds.) *Cretaceous carbonates of Texas and Mexico: applications to subsurface exploration, Rep. Invest. Bur. econ. Geol. Univ. Texas.* **89**, 325–331.

Young, R.G. (1960). Dakota Group of Colorado plateau. *Bull. Am. Assoc. Petrol. Geol.* **44** (2), 156–194.

Yun Fu Zeng, Nan Hao Lee and Yang Zhou Huang (1983). Sedimentary characteristics of oolitic carbonates from the Jialing-Jiang Formation (Lower Triassic, T Jl), South Sichuan Basin, China. In: Peryt, J.M. (ed.) *Coated Grains.* Springer-Verlag, Berlin, 176–187.

Zacher, W. (1973). Das Helvetikum zwischen Rhein und Iller (Allgäu-Vorarlberg): Tektonische, paläogeographische und sedimentologische Untersuchungen. *Geotekton. Forsch.* **44**, 74pp.

Zambrano, E., Vasquez, E., Duval, B., Latreille, M. and Coffinieres, B. (1972). *Paleogeographic and petroleum synthesis of western Venezuela.* Editions Technip. Paris, 67pp (bilingual).

Zankl, H. (1971). Upper Triassic carbonate facies in the northern Limestone Alps. In: Muller, G. (ed.) *Sedimentology of parts of Central Europe, Guidebook to excursions. 8th Int. Sedimentol. Congr., Heidelburg, 1971.* Verlag Waldemar Kramer, Frankfurt, 147–186.

Zapp, A.D. and Cobban, W.A. (1960). Some Late Cretaceous strand lines in northwestern Colorado and northeastern Utah. *Prof. pap. U.S. geol. Surv.* **400-B**, 246–249.

Zapp, A.D. and Cobban, W.A. (1962). Some Late Cretaceous strand lines in southern Wyoming. *Prof. pap. U.S. geol. Surv.* **450-D**, 52–55.

Zeiss, A. (1977). Jurassic stratigraphy of Franconia. *Beitr. Naturkd. Ser. B.* **31**, 1–32.

Ziegler, A.M., Scotese, C.R., Bambach, R.K. and Parrish, J.T. (1980). *Triassic and Jurassic paleogeography of China.* Unpubl. Ms. Univ. Chicago.

Ziegler, M.A. (1967). A study of the Lower Cretaceous facies developments in the Helvetic Border Chain, north of the lake of Thun (Switzerland). *Eclog. geol. Helv.* **60** (2), 509–527.

Ziegler, P.A. (1969). The development of sedimentary basins in Western and Arctic Canada. *Excursion guidebook for Int. Devonian Symp., Calgary 1967; 2nd Ed., Alberta Soc. Petrol. Geol., Calgary.* 95pp.

Ziegler, P.A. (1975). North Sea basin history in the tectonic framework of N.W. Europe. In: Woodland, A.W. (ed.) *Petroleum and the Continental shelf of North West Europe,* **1**, *Geology.* Applied Science Publishers, London, 131–148.

Ziegler, P.A. (1977). Geology and hydrocarbon provinces on the North Sea. *Geojournal.* **1** (1), 7–32.

Ziegler, P.A. (1978). North-Western Europe: tectonics and basin development. In: Loon, A.J. van (ed.) *Key-notes of the MEGS-II, Amsterdam, 1978, Geol. Mijnb.* **57** (4), 589–626.

Ziegler, P.A. (1980). Northwest European basin: geology and hydrocarbon provinces. In: Miall, A.D. (ed.) *Facts and principles of World petroleum occurrence. Mem. Can. Soc. Petrol. Geol.* **6**, 653–706.

Ziegler, P.A. (1981). Evolution of sedimentary basins in North-West Europe. In: Illing, L.V. and Hobson, G.D. (eds.) *Petroleum geology of the continental shelf of north-west Europe. Proc. 2nd Confr., London, March 1980, Inst. Petrol.* Heydon & Son, London, 3–39.

Ziegler, P.A. (1982a). *Geological Atlas of Western and Central Europe.* Shell International Petroleum Maatschappij B.V., The Hague, 130pp, 34 maps, 4 correlation charts, 2 legends.

Ziegler, P.A. (1982b). Triassic rifts and facies patterns in Western and Central Europe. In: *Proc. Symp. Triassic global geology, Wurzburg, Feb. 1982, Geol. Rdsch.* **71** (3), 747–772.

Ziegler, W.H. (1975). Outline of the geological history of the North Sea. In: Woodland, A.W. (ed.) *Petroleum and the Continental Shelf of North West Europe,* **1**, *Geology.* Applied Science Publishers, London, 165–187.

Zilli, N., Orchuela, I., Dellape, D. and Otano, R. (1978). Análisis de las Formaciones Quintuco y Loma Montosa en el sector centro oriental de la Cuenca Neuquina. In: *Acta 7th Congr. Geol. Argentina, Neuquén, April 1978, I, Sect III, Geologia histórica y regional.* 609–615.

Zinsmeister, W.J. (1982). Late Cretaceous-Early Tertiary molluscan biogeography of the southern circum-Pacific. *J. Paleontol.* **56** (1), 84–102.

Zi-shun, L. (1980). The stratigraphic regionalization of the Triassic in China. In: Gaetani, M. (ed.) *Contributions to the Triassic stratigraphy, Proc. Riccardo Assereto and Giulio Pisa field Symp. on Triassic Stratigraphy of the Southern Alps, Bergamo, June 1979. Riv. Ital. Paleontol. Stratigr.* **85** (3–4), 1147–1156.

Znosko, J. (1968). Prog Pompeckiego i jego paleogeograficzne znaczenie dla polskiej jury. *Kwart. geol.* **12** (2), 316–323.

Znosko, J. (1968). Ozcylacje transgresywne morza doggerskiego między Górzowem Wielkopolskim i Zakrzewem. *Kwart. geol.* **12** (2), 308–315.

Printed in the United States
By Bookmasters